The
Golden Age
of
Beekeeping

Peter Loring Borst

Northern Bee Books

The Golden Age of Beekeeping

ISBN 978-1-914934-82-7

Published by Northern Bee books, 2024
Scout Bottom Farm
Mytholmroyd
Hebden Bridge HX7 5JS (UK)

Design and artwork by DM Design and Print

The
Golden Age
of
Beekeeping

Peter Loring Borst

Contents

Introduction

When I first got the idea to write this story, I was intending to call it the "History of Beekeeping in New York." But my wife, who is perhaps more realistic than I, stated that such a book would be of interest to about fifty bald guys with beards. This took most of the wind out of my sails, but a new gust blew through and I decided to affix a new title, which she thought a great improvement. The truth is, the *Golden Age of Beekeeping* did take place largely in New York State.

The reasons for this are many, but among them include the vast areas planted to clover and buckwheat and the huge population centers in and around New York City. Yet, more than that, it was many ingenious and forward looking New Yorkers who took beekeeping from a quaint cottage industry and built it into a profitable enterprise. For the sake of comparison, let's look quickly at some figures. In 1860, New York produced 2.5 million pounds of honey, and 138 thousand pounds of beeswax. In 2011, the honey production in NY was almost the same, 2.8 million pounds. But bear in mind, in 1860 this was all done with man power and horse drawn carriages. Produce was shipped downstate via barges.

But I am getting ahead of myself. To tell the story properly, we must go back to the time when the so-called New World was being colonized by Europeans. They found the country filled with possibilities but vastly unfamiliar. Lacking were most of the plants and animals they were accustomed to. Additionally, the country was already populated by human beings whose ways seemed primitive and obsolete. And yet, these very people help the settlers adjust to the land and its resources. All the same, the change had begun: Europeans were determined to bring their culture, with its livestock and crops, weeds and pests and all, to the Americas.

One of these was the honey bee, which was nowhere to be found. We often hear that they did this to provide pollinators for their crops. The fact is, there were plenty of pollinators here already, in the form of native bees, wasps, flies, beetles and so on. It wasn't until centuries later that the role of pollinators was even understood. The settlers brought bees because they could provide a concentrated sweetener, to alleviate the monotony of a diet of meat, grains, and what vegetables they could raise.

In this book I tell the story not only of the bringing of honey bees to America, but also how beekeeping changed from a minor accompaniment to the family's garden, to a significant industry more like dairy farming. The beekeeper went from being a slightly odd character having some secret knowledge and a bit of tolerance for bee stings—to a major partner in the industrial revolution.

All of this took a few of centuries. Strictly speaking, though, the Golden Age is the period from 1865 when *The American Bee Journal* began its unbroken print run (which continues to this day), to about 1920, when the use of motorized cars and trucks changed beekeeping forever.

Acknowledgements

A work like this could not come about without the help of many people. I apologize in advance to whomever I fail to mention. Professor Thomas Seeley provided the initial impetus to do this book, having read my articles over the years. Eugene Makovec, the editor at *The American Bee Journal*, has given guidance, support, and encouragement without which I would not have gotten this far. This entire book was published in a different form as a series of articles in the journal. My personal editor and critic is my wife Diane. She always makes certain that the final product is clear, readable and fun. Finally, my friend Christel Truttman provided valuable proofreading and editing assistance as we approached the final draft. I confess we all feel that much more could have been included; perhaps a sequel will emerge. Thanks for reading.

Peter Loring Borst
Danby, NY USA
March 2024

Chapter 1

Coming To America

He who rescues from oblivion interesting historical facts is beneficial to posterity as well as to his contemporaries, and the prospect thereof to a benevolent mind causes that employment to be agreeable and pleasant which otherwise would be irksome and painful (Hutchinson 1769).

Beginnings

The story of the honey bee in North America is based on hints and interpretations, larded with streaks of hyperbole. By the 1700s, honey bees were so numerous the question arose as to whether they were actually native to America. Benjamin Smith Barton treated it extensively in 1793, in his "An Inquiry into the Question, whether *Apis mellifera*, or True Honey-Bee, is a Native of America." He wrote:

> *So many animals and vegetables have been introduced into the countries of America, since the great discovery of Columbus, that naturalists are frequently at a loss to determine, which species are natives, and which are foreigners. (Barton 1793)*

Barton lists some of the introduced plants, including plantain (*Plantago*), mullein (*Verbascum*), lamb's quarters (*Chenopodium*), St. John's wort (*Hypericum*), and the common dandelion (*Taraxacum*). Of animals, he mentions rats, mice, clothes moths, fleas, and bedbugs. Probably only the moths were new; the New World was amply supplied with rats, mice and the rest. Europeans definitely brought diseases with them, including smallpox, typhus, measles, chicken pox, mumps, and the flu. But what about honey bees?

Barton cites work by his contemporary, Dr. Jeremy Belknap, who wrote in depth about the travels of Christopher Columbus. Appended to his work, he also questioned whether the honey bee is native and lists many reasons to suggest that it is not. Belknap wrote:

> *There is a tradition in New England, that the person who first brought a hive of bees into the country was rewarded with a grant of land;*

but the person's name, or the place where the land lay, or by whom the grant was made, I have not been able to learn. (Belknap 1792)

ₐ"SMUDGING FOR BEES." "WORKING DOWN THE BEE-TREE."
BEE HUNTING IN THE ADIRONDACKS.—[Sketched by Theodore R. Davis.]

Bee Hunting

Belknap refers to accounts of the Spanish, who exploring Mexico, found the natives producing honey. Stingless bees of various types (genera *Melipona, Trigona, etc.*) are found throughout tropical America; none of them are true honey bees (genus *Apis*). These are undoubtedly what they encountered. As evidence of this, we can take the mention of a tribute being paid with 600 cups of honey, which is a particularly small amount compared to the yield of true honey bees. The harvest from a stingless bee nest can be measured in cups; the yield of a honey bee hive can be many gallons. Further, Belknap mentions the wax and said: "Though they extracted a great quantity of wax from the honey comb; they either did not know how, or were not at the pains to make lights of it."

What he means is that they didn't produce candles. Unlike honey bees, which build comb from pure beeswax, stingless bees form their rudimentary cells from a substance called *cerumen*, also called "black beeswax." It contains small amounts of wax secreted by the bees, but it is primarily made from things they gather such as plant resins and also soil, seeds, even small stones. Hence it would be hardly suitable for candles. Native Amazonians use it as an adhesive. They cook it, which causes it to blacken, and use it in making arrows. Still, as late as 1809, writers were adamant that honey bees had been here all along. Samuel Williams wrote in his "History of Vermont:"

> *Whether the honey bee is a native of the country, seems to be viewed by some as uncertain. I do not find much reason to doubt but that it was in America, before the Europeans made their first settlements in the country. They live in the hollow trees in the woods of Vermont, from year to year; and are always found, of their full dimensions, vigorous, and plentifully supplied with honey; and they bear the cold of our winters, much better in the hollow of a large tree, than in any of our artificial bee hives. (Williams 1809)*

Bees in the Woods

Honey bees were indeed abundant in the woods, as evidenced by stories about bee hunting, with instructions on how to find them using the time-honored technique of "bee lining" or following bees back to their nest. Paul Dudley, a lawyer born in Massachusetts in 1675, was an accomplished naturalist. He was made a Fellow of the Royal Society of London in 1721. The Royal Society had been founded in 1660 and published scientific articles on a wide range of topics from astronomy to biology. It printed several of Dudley's articles, including one where he described and illustrated the method of discovering bee-trees and how to rob the honey:

> *All the Bees we have in our Gardens, or in our Woods, and which now are in great numbers, are the produce of such as were brought in Hives from England near a hundred Years ago, and not the natural produce of this part of America; for the first Planters of New England never observed a Bee in the Woods, until many Years after the Country was settled; but that which proves it beyond question is, that the Aborigines (the Indians) have no word in their Language for a Bee, as they have for all Animals whatsoever proper to, or aboriginally of the Country. (Dudley 1721)*

This example serves to illustrate several important facts. First of all, by the 1700s, honey bees were numerous in the forests surrounding the settlements. Second, beekeeping itself was not really the main source of honey; it was more common to obtain honey by robbing feral bees. Dudley may have been the first to write about it, but many others did as well.

ROBBING A WILD-BEE HIVE.—Drawn by R. F. Zogbaum.

Robbing Wild Bees

Henry Rowe Schoolcraft was born March 28, 1793 in Albany County, New York and led several expeditions to what was then the "wild west." He published a book titled *Rude Pursuits and Rugged Peaks: Schoolcraft's Ozark Journal, 1818-1819*. In it, he gave a description of bee hunting in the woods:

> We discovered a bee-tree, which Mr. Pettibone and myself chopped down. It was a large white oak, (Quercus alba) two and a half feet across at the butt, and contained, in a hollow limb, several gallons of honey. This was the first discovery of wild honey which accident had thrown in our way. It should here be remarked, that the white hunters in this region (and I am informed it is the same with the Indians) are passionately fond of wild honey, and whenever a tree containing it is found, it is the custom to assemble around it, and feast, even to a surfeit. (Schoolcraft 1821)

James Fenimore Cooper penned a novel called *The Bee Hunter*, whose main character he named "Ben Buzz." Cooper says:

> As he was one of the first to exercise his craft in that portion of the country, so was he infinitely the most skilful and prosperous. There were a score of respectable families on the two banks of the Detroit, who never purchased of any one else. (Cooper 1848)

In the literature of the 1600s, there are references to bee hives in the American Colonies, but many of these are of doubtful authenticity. I will give one example:

> What shall I say more? you shall scarce see a house, but the South side is begirt with Hives of Bees, which increase after an incredible manner: That I must needs say, that if there be any terrestrial Canaan, 'tis surely here, where the Land floweth with milk and honey. (Denton 1670)

Unfortunately, Mr. Denton was prone to exaggerate, presumably to entice travelers to populate the struggling colonies. He claims in the same paragraph that "many people in twenty years time never know what sickness is," and that as few as two or three deaths occurred annually in a town. According to John Duffy, in his book on the history of public health of New York City from 1635 on:

> A word of caution should be given here. In the seventeenth and eighteenth centuries the New World was rightly looked upon as an

> *exotic and marvelous place, and travel writers who wished to keep*
> *their readers dared not disturb this image. Moreover, many works*
> *were written under the patronage of wealthy land grant holders who*
> *were eager to people their vast holdings in America. (Duffy 1966)*

Other writers, instead of arguing if or when bees were imported to the coast of America, tracked its progress across the wilderness. Gene Kritsky, professor of Biology at Mount St. Joseph University and the editor of *American Entomologist,* has been writing about honey bees for decades. His 1991 article in the *American Bee Journal* has detailed maps showing the years of first sightings of honey bees which had escaped into the wild, beginning with their arrival at Jamestown, Virginia in 1622, on to being seen in Georgia in 1736, a distance of about 500 miles. At this rate they would be traveling about 5 miles per year. Kritsky has them at the Mississippi River by 1800, based upon various sources.

It seems clear from the evidence that more honey was obtained from wild bees living in hollow trees, than was harvested from what manmade hives may have been kept in the new settlements. As further proof of this suggestion, I found mention of the subject in an early issue of the *Scientific American* magazine, which had been established in 1845 in New York City. This paragraph written in 1897 gives a glimpse of beekeeping of the past as well as the late 1800s:

> *Half a century ago honey was considered a luxury, and the market*
> *was supplied by the professional bee hunters, who made a*
> *precarious living in locating the natural hives of the bees in some old*
> *rotten tree right in the midst of the thick forest; but to-day, 30,000*
> *bee keepers vie with each other to supply us with all the varieties of*
> *delicious honey that we are willing to pay for, and at prices within the*
> *reach of every one. (Walsh 1897)*

Overseas Journeys

In the course of trying to find out as much as possible about how and when honey bees were brought to colonial America, I found a comment in a book about the ocean voyages and the ships that were used. The author wisely stated that most of the details were so commonplace that they didn't merit mention at the time. On the other hand, a detailed description of a typical sailing ship can give us a general description which would apply to the others.

The Mayflower had been used for years to transport wine before it was chartered to carry pilgrims to the New World; it was therefore called a "sweet ship" in contrast to ships that routinely carried more mundane cargo. Crossing the Atlantic in the 1600s was not without perils, but the ships themselves were rarely the cause of death; only a few were wrecked. Most often malnutrition and disease were the greatest hazards. The passengers were generally treated little better than livestock, which the ships usually carried as well. Animals such as cows, pigs, dogs and the like were very much in demand in the colonies and would fetch a high price; some of them were of course eaten by the voyagers. The captain generally feasted the entire voyage, while the passengers were given the cheapest food possible: hard biscuits, pickled meat, and beer barely fit to drink.

> The problem of food and water on these long passages was serious; Cotton Mather relates several instances of vessels which exhausted their supplies, and the crews of which were miraculously delivered from starvation. A small Vessel set sail from Bristol to New England, September 22, 1681, with the Master, whose name was William Dutten; there were seven Men a board, having Provisions for three months. "She was delayed so long at sea by headwinds and bad weather that all her supplies were consumed; the beer was exhausted, and most of the drinking-water was lost by reason of leaky casks. The ship's company lived for a time on rats and rain-water "drinking a thimblefull at a time." (McElroy 1935)

Contemporary ship logs and bills of lading show that settlers attempted to bring bee hives with them on several occasions, but there are no records showing whether they survived the trip or the early years of settlement when starvation was the greatest peril the newcomers had to face.

> Seventeenth-century merchant vessels, bluff ended and round bottomed, were slow and inefficient sailers that bobbed like apples in wind-whipped sea swells. Every inch of cargo space was precious; so a skipper jammed colonists, belongings, livestock, and freight together in his ship's small hold. Disease was common, since sanitary facilities were minimal. Fresh water was scarce, and even cooking hot meals invited the risk of setting the ship afire. (Billings 2012)

The Mayflower anchored at Plymouth Rock in late 1620, and formed the first permanent settlement of Europeans in New England. More than half of the original settlers died during that grueling first winter. Since so many people as well as their livestock perished, it is fairly reasonable to suppose that many of the bees also did not survive. Eventually, though, live bees escaped into the woods surrounding the colonies at Plymouth and Jamestown, as evidenced by ample records of bee hunting in the colonies.

Mayflower

Early References

This is the earliest reference to a shipment of bees to Virginia:

> *Wee have by this Shipp and the Discoverie sent you divers sorts of seeds and fruit trees as also Pidgeons, Connies [rabbits], Peacocks, Mastives and Beehives, as you shall by the Invoice perceive. - Letter to the Governor and Council Of Virginia, dated Dec. 5, 1621. (Neill 1869)*

If you consult the very many books and articles written on the history of beekeeping, you will often see the date 1638 cited as when honey bees were successfully brought to New England. But as Lee Watkins tells in his piece entitled "First Honey Bees in New England – 1638?" (*American Bee Journal*, 1968) this is a perennially repeated error based on an inaccurate reading of "An Account of the Voyages to New England" by John Josselyn, published in London in 1674. Josselyn states plainly that honey-bees were carried over by the English, and "thrive there exceedingly!" – but he does not affix a date to their arrival.

I often wondered how the bees were transported. William Cotton describes an elaborate plan to transport bee hives to New Zealand. His scheme was to have the bees placed on ice in the ship's bilge, which would be the coldest, darkest place. However, the bilge was also the most unhealthy place in the ship, where foul water and other waste accumulated, so I doubt this plan was carried out. A more realistic approach was described by Edward Goodell, in his article titled "Bees by Sailing Ship and Covered Wagon." He claims to have found a book published in Antwerp in 1830, but offered no further bibliographic information. Goodell describes in great detail how hives would be shipped:

> *A strong crate or chest was built to hold the hives. (I presume they were straw skeps.) This crate had two shelves built into it, one above the other then was divided into four compartments into which each of the skeps fitted snugly. (Goodell 1969)*

The entrances of hives were closed while placed on board, and kept closed till they were out at sea. According to Goodell, a platform was built at the stern of the ship and the crate with the hives was bolted to it. The crate had ventilating holes in the bottom to allow air flow. They could be shut if the weather was very bad but also opened to allow the bees to come out and fly. The bees were unlikely to wander far from the ship, as there were no flowers to entice them, and they would easily find their way back: bees naturally memorize their surroundings and the only thing memorable at sea would be the ship. Goodell says that the shippers knew that the bees would not do well if kept below deck for many months, so they were kept topside. Many of these ships carried a variety of animals, as I already mentioned.

An interesting parallel exists between the importation of bees and of other livestock. G. A. Bowling, writing in 1947 in the *Journal of Dairy Science* states that "The scarcity of data relative to the first importations of cattle into Colonial North America has lent obscurity to one of the most interesting phases of early American husbandry." So the lack of concrete evidence is not unique to the story of the honey bee.

Uncle Dan Myers and bees

Some Remarks

Despite the fact that the record is littered with gaps, inaccuracies and fiction, I think there are several conclusions that can be gleaned. The colonization of the American colonies took place over many years, and was accompanied by many failures, caused by lack of preparedness due to insufficient resources, poor planning, and plain bad luck.

While it is certain that honey was valued by the settlers, in isn't clear how much skill or attention was directed toward bees, at least in the beginning. Perhaps honey bees were much more capable of making do with what was available than the European colonists, who relied very heavily on trade with and instruction from the native "Indians." Another researcher came to much the same conclusion:

> So prevalent had the forest bees become that by the late seventeenth century most people could rely on "wild" honey for their sweetening, rather than actively cultivating bees. It was considered a challenging and dangerous sport, as well as a productive activity – a bee tree often produced as much as fifty to seventy-five pounds of honey. The availability of wild honey held back the growth of formal apiculture in the Chesapeake region. (Pryor, 1983)

In summation, honey bees as well as numerous other species escaped and found niches in the New World. Mostly, they simply found a way to "fit in." In many cases they became what are called "invasive species," which often crowd out the locals using traits such as rapid growth and reproduction, wide dispersal, and adaptability. In this sense, both the European settlers and honey bees invaded the American continent and now are found almost everywhere.

Works Cited

Barton, B. S. (1793). An inquiry into the question, whether the Apis mellifica, or true Honey-bee, is a native of America. Transactions of the American Philosophical Society, 3, 241-261.

Belknap, J. (1792). A Discourse, intended to commemorate the Discovery of America by Christopher Columbus. Printed at the Apollo Press, in Boston, by Belknap and Hall, State Street.

Billings, W. M. (2012). The Old Dominion in the Seventeenth Century: A Documentary History of Virginia, 1606-1700. UNC Press Books.

Dudley, P. (1721). An account of a method lately found out in New-England, for discovering where the bees hive in the woods, in order to get their honey. Philosophical Transactions of the Royal Society of London, 31(367), 148-150.

Duffy, J. (1968). History of public health in New York City, 1625-1866. Vol. 1. Russell Sage Foundation.

Goodell, E. (1969). Bees by Sailing Ship and Covered Wagon. GLEANINGS IN BEE CULTURE, 97(1), 38.

Hutchinson (1769). A Collection of Original Papers Relative to the History of the Colony of Massachusetts-Bay. New York: Burt Franklin.

McElroy, J. W. (1935). SEAFARING IN EARLY NEW ENGLAND. New England Quarterly, 8(1), 331.

Neill, E. D. (1869). History of the Virginia Company of London: With Letters to and from the First Colony, Never Before Printed. Joel Munsell.

Pryor, E. B. (1983). Honey, Maple Sugar and Other Farm Produced Sweetners [sic] in the Colonial Chesapeake. Accokeek, Maryland: The National Colonial Farm Research Report.

Schoolcraft, H. R. (1821). Narrative journal of travels through the northwestern regions of the United States. E. & E. Hosford.

WALSH, G. E. (1897). Honey and Bee Keeping. Scientific American, 77(11), 167-167.
Williams, S. (1809). The natural and civil history of Vermont (Vol. 1). Samuel Mills.

Chapter 2

The Slow Advance of Beekeeping, 1600-1800

Tammy Horn, in her excellent book Bees in America: *How the Honey Bee Shaped a Nation*, asserts that "When the English took over the New Netherlands in 1664, beekeeping was already established in this colony." Dr. Horn's book is thorough in its documentation of history, but in this case there is no support. I have been combing through the history of New Netherlands (which became New York) for many years to find records of beekeeping prior to the 1800s, and have turned up next to nothing.

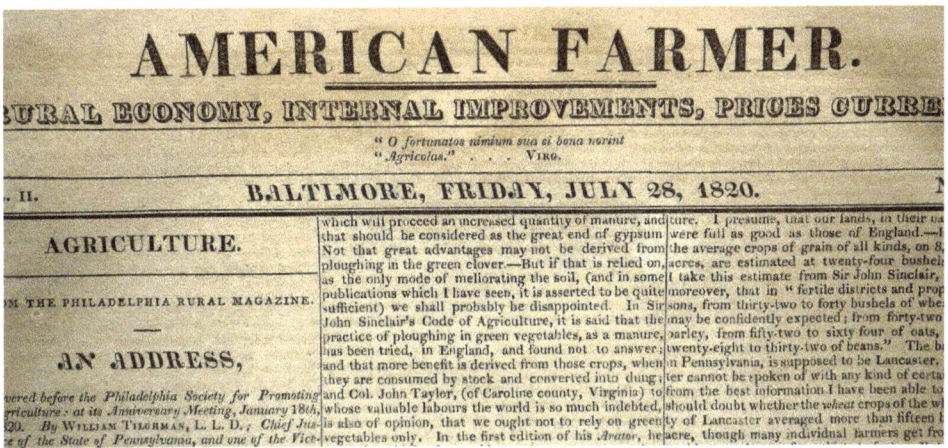

American Farmer 1820

We don't really know who brought the bees, and what they had in mind. Probably they were brought by settlers, or enterprising shippers hoping to sell them to settlers. The evidence is simply not there to support a clear picture. On the other hand, much is known about the introduction of other livestock, such as cattle, horses, and pigs. The absence of clear records of beekeeping in the colonies is therefore puzzling, but could certainly indicate a state of benign neglect.

What is clear is that the hives brought to the New World swarmed and the bees headed to the woods. So prevalent had the forest bees become that by the late seventeenth century most people could rely on "wild" honey for their sweetening, rather than actively cultivating bees. Writing in her book about life in the American Colonies in the 1600s, Elizabeth Brown Prior says that bee hunting was a popular pastime.

Beekeeping in the Old Country

By the late 1600s, a lot had been printed on the technique of keeping bees. But, we don't really know how widely these publications were read. Books were expensive in the colonies and reading non-religious content was discouraged.

> *It would be worth having a full-scale bibliographical study of science in America, but the general picture seems to be clear: that down to the latter part of the nineteenth century American science can best be seen as a provincial part of British science, and American books usually reprints, authorized or pirated, of those originating in this country [England]. (Knight 1989)*

In England, beekeeping was well advanced, as evidenced by the bounty of literature of the times. Foremost among the important books of the time is Butler's *Feminine Monarchie*, first published in England, in 1609. Butler was not the first to refer to the queen bee (as opposed to the centuries old error of the hive being ruled by a "king") but he made it central to his work, as evidenced by the title, *Feminine Monarchie*. He describes her like this (rendered in modern English):

> *The Queen is a fair and stately bee, differing from the common both in shape and color: her back is all over of a brighter brown: her belly even from the top of her fangs, to the tip of her train, is of a sad yellow, somewhat deeper than the richest gold. She is longer than a honey-bee, almost an inch long: she is also bigger than a honey-bee, but not so big as a drone ... The spear she has is but little, and not half so long as the other bees: which, like a King's sword, is born rather for show and authority, than for any other use.*

H. Malcolm Fraser, the great British beekeeping author and researcher, wrote that upon opening the book "one seems to enter a new world." It has a readable straightforward narrative, and Fraser wrote in 1956 that there was still no better English work on skep beekeeping. He continues: "The book does not only deal with practical skep management. It is a treatise on general beekeeping, and could even now be read with profit by a beginner."

Beekeeping was widely practiced in the 1600s in Northern Europe. The bees were kept mostly in *skeps* (basket-like hives made of wicker or straw; from Old Norse *skeppa* = basket, bushel; Gaelic *sgeip* = a beehive). Also, hollow log hives were used and some beekeepers were starting to use wooden boxes.

PURE HONEY of the following SORTS,
In the greateſt PERFECTION,

At *Wildman's* Bee & Honey Warehouſe,

Nº. 326, HOLBORN, almoſt oppoſite *Gray's-Inn-Gate.*

FINE Virgin Honey at 1s. per Pound, and a particular Sort of fine pure Virgin Honey, gathered in the Spring, and extracted from the fineſt Combs, in the moſt delicate Manner at 1s. 6d.
per Pound.—This Honey is recommended by the moſt eminent Phyſicians for ſpreading on Toaſt,
ſweetening Tea, &c. It preſerves the Lungs, and prevents many of the worſt Diſorders; and,
when properly taken, is a certain Cure for ſeveral others, particularly Coughs, Hoarſeneſs, Aſthmas,
Conſumptions, and the Gravel.—Fine Virgin Honeycomb, made in his Glaſs Hives, at 2s. 6d.
per Pound, not to be equalled in this Country.—Virgin Honey clarified at 2s. per Pound; Honey
proper for making Mead from 6d. to 8d. per Pound; real Minorca, Narbonne, and other Foreign
Honey juſt imported.—Druggiſts, Apothecaries, Grocers, &c. may always be ſupplied with Honey
Wholeſale, and any Quantity may be had for Exportation.

Alſo, a great VARIETY *of his*

New-invented Mahogany, Glaſs *and* Straw Bee-Hives,

(Both for C H A M B E R and G A R D E N,)

So much approved of by the Nobility, Gentry, and others; and by which the Lives of thoſe curious
and uſeful Inſects, the BEES, are ſaved Theſe Hives afford an Opportunity of taking a treble
Quantity of Honey to any other Hives whatever; the Bees may be ſeen at work in their different
Apartments, and their Honey taken in the greateſt Perfection at Pleaſure: And as Mr. WILDMAN
keeps a great Number of Bees in different Parts of the Country, Orders for any Quantity of them
will be ſupplied on the ſhorteſt Notice.

Juſt publiſhed, and may be had as above, Price 1s. 6d. *illuſtrated with Copper-plates,*

His Complete Guide for the Management of B E E S throughout the Year.

The FOURTH EDITION, with ADDITIONS.

☞ *It likewiſe may be had tranſlated into* F R E N C H.

Mr. WILDMAN returns his moſt grateful Thanks to the Nobility and Public in general, for
the great Encouragement he has met with for many Years paſt, and is happy to find his peculiar
Method of taming and cultivating BEES has enabled him to ſell his beſt Honey as above, which is
far ſuperior in Purity and Flavour to any yet offered to the Public; and at the ſame Time aſſures
them, that he ſhall ever make it his Study to preſerve that decided Preference in his Favour which he
has hitherto had the Pleaſure to obtain.

Wildman's Bee _ Honey Warehouse 1770s

Keeping Bees in Skeps

Skep beekeeping consists principally of capturing free flying swarms by collecting or enticing them into the skep. Once a swarm of bees is clustered, say: hanging within reach on the branch of a tree as they often do, a skilled beekeeper can shake the cluster into the hive and set it down. Thus, the bees find themselves in a dry dark cavity, which is what they naturally prefer for a dwelling place. The hive would necessarily be built to a particular size, learned through trial and error, and often the interior surface would be rubbed with herbs or beeswax as a further enticement. Sometimes honey bee swarms reoccupy hives in which the prior tenants perished; a well-used bee hive is almost irresistible to an itinerant swarm.

The goal is to catch swarms of bees in late spring or early summer; there is an old proverb: "A swarm in May is worth a load of hay." This, because they would store up a supply of honey which the beekeeper would then harvest, either by the humane method of removing only part of it so that they should survive, or else by suffocating the bees and taking all of their honey. The basic skills in skep beekeeper include being able to tell if a given bee colony is about to swarm, and the stealing of their honey.

The major advance was the addition of compartments either above or below the skep hive in order to increase the space available to the bees which they often would immediately occupy. This led to the technique of separating the sections so that one might take part of the honey without having to use the messy technique of cutting the honey combs while they were still covered with bees.

By removing the added section, or "eke" as it was called, and placing it away from the main hive, the bees would abandon it and return to the nest. If the queen happened to be in the removed section, the bees would not abandon her, and tended to stay with that part. Observant beekeepers learned that a hive which lost its queen would raise another one, using very young bee larvae, fed a special diet called "royal jelly." Substances in this food cause the larva to develop into a queen honey bee, rather than a regular "worker bee" which it would have done. These discoveries led to the practice of dividing hives at will to increase the number of bee colonies, instead of depending on the vagaries of natural swarming.

Beekeeping at the time consisted of two distinct and separate streams. The small holder, or cottager, as they were called, kept bees according to tradition and was slow to innovate. Meanwhile, inquisitive practitioners were pursuing the burgeoning fields of science, using new tools such as the microscope and engaging in methodical experimentation to determine the principles upon which the field of natural science would be constructed. Dr. H. Malcolm Fraser covers the progression of beekeeping literature in his 1958 book, *The History of Beekeeping in Britain*. He cites Butler's *The Feminine Monarchie* as a milestone in beekeeping literature. One of the themes that runs through the literature, from William Lawson's 1617 *A New Orchard and Garden* and *The Country Housewife's Garden* on through its successors such as the Reverend William Cotton's *Letter to Cottagers* was the encouragement of small holders. The keeping of bees was and still is compatible with small scale, intensive agricultural pursuits; requiring little capital but depending on labor available on the farm. Following the techniques gleaned from centuries of practice, bees could be exploited for their treasure of honey and wax. Large scale beekeeping had not been contemplated in the States, even though there were many beekeepers in Europe and elsewhere that had apiaries containing hundreds of hives.

Lexden Pocock (1850-1919)

Beekeeping in the U.S.

Beekeeping in the U. S. can not be said to have advanced very much by the late 1700s, despite the mentions it gets in various publications. Bee hunting was a favorite pastime; beekeeping was not. In 1787, the following appeared in one of the first newspapers printed in the newly formed country, namely *The New-Hampshire Spy*.

> *Many and great are the advantages to be gained by the inhabitants of these United States, if bees were propagated, supported, and preserved. Our soil and climate are inferior to none for this purpose. Canaan, of old, could not with more propriety be called a land flowing with milk and honey, than America would be, did we but improve all the means to produce these so valuable and so important articles, which we might do very easily; which would assist each other when we annually extended such pastures as would increase both. (New-Hampshire Spy 1787)*

Here is a significant change in the rhetoric. In the beginning, creative writers said the new world is the land of milk and honey; here the author is stating that it could be, if honey bees were propagated, supported and preserved.
Not simply destroyed for their honey but cultivated as livestock. Small quantities of sweets could be gleaned from boiling down maple sap, or apple cider, or by robbing the bee-trees in the woods. Whereas, very large amounts of honey and beeswax would be obtained by greatly increasing the number of kept hive, and by discontinuing the practice of killing bees for their honeycombs.

Philip A. Mason has diligently compiled a bibliography of all of the books on bees published in English in the United States and Canada. His work is an excellent resource for anyone interested in the progress of apiculture in the Americas. He telSl us that Thaddeus Minor, of Woodbury, CT wrote one of the earliest, quite possibly the first, original work on beekeeping in the New World, which was published in 1804. According to Mason, the Library of Congress has a copy of it. I was able to obtain a "microprint" of the work from the Cornell University library. So far as I know, few people have ever seen this work.

The title is epic, and summarizes the work; the use of an extended title in this manner was common at the time.

The Experienced Bee-Keeper; or, a Short Treatise on the Management of Bees; Founded on Facts and long Experience: Wherein is Described, Their nature and kind, and how to manage them. When they swarm, how to place them. How to take care of them through the winter. How to feed them. How to destroy their enemies. How to unite swarms. How to manage them when robbers attack them. How to form a double hive. How to preserve honey; together with some of its qualities and uses. (Minor 1804)

Minor quite obviously knew a lot and had experience keeping bees. He recommended making hives out of hollow logs, naming pine the best, though chestnut or oak would do. He told his readers that the logs must be clean and that a board must be tightly nailed to the top to prevent water from leaking in, which he cited as fatal to the bees in winter. Finally, he said: "Straw hives I have never used, and therefore shall have nothing to say about them." It is not clear that straw hives were used at all in the Americas. Bees were probably brought across the Atlantic in traditional straw skeps. However, we don't know what became of them. Unless there were individuals skilled in the making of skeps, it's plausible that the bees swarmed out of them and the skep hives themselves were not replaced, since straw hives are not durable.

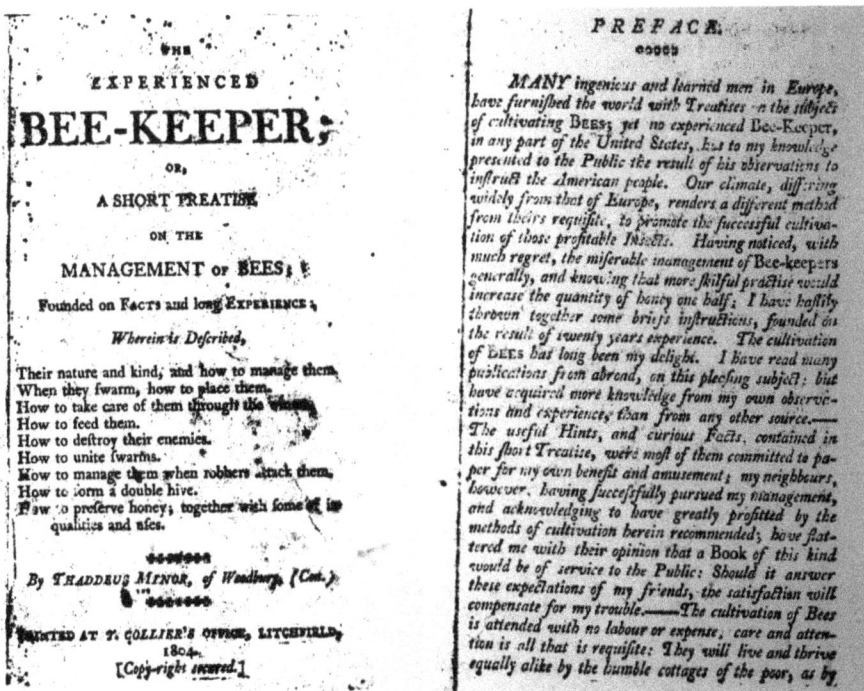

The experienced bee-keeper 1804

Mr. Minor understood well enough that a large amount of honey could be gotten from a hive of bees. He wrote of one hive that weighed thirty pounds in spring, reaching two hundred in the summer. Oddly, he said nothing of harvesting the honey. The usual way of doing so was to hold the hive over a pit of burning sulfur, causing the bees to asphyxiate, die, and leave the owner with a basket or log full of honey combs.

Minor was acquainted with the externalities of bee hive management but alas, he was painfully ignorant of the internal workings. He told readers that there are three different sorts of bees in the hive (true enough). He correctly identified the "common workers" and "the drones" but erroneously referred to the drones as females. He noted correctly that they do no work, but suggested they contribute to the care of the young brood, if only by helping to warm it. Where he really fell down was in his description of the third type of bee:

> The King Bee is very different from the rest, both in shape and colour. His body is longer than the Drone, neater made, and tapers to a point. He has very short wings in proportion to his size. (Minor 1804)

When Mr. Minor published this, almost 200 years had elapsed since Charles Butler wrote his masterwork *The Feminine Monarchie*, where he declared the big bee "The Queen." And even a dictionary from Minor's period states:

> Queen-Bee, a name given by late writers to what used to be called the king-bee, or king of the bees; a large and long-bodied bee, of which kind there is only one found in every swarm. (Croker 1766)

Thaddeus Minor, despite publishing in 1804, belonged securely to the ways of the 1700s. H. M. Fraser wrote:

> The opening of the nineteenth century may conveniently be regarded as a turning point in the history of beekeeping. Until 1800, science and practice in beekeeping had been considered completely distinct and unconnected. (Fraser 1957)

Before describing the progress of practical beekeeping in France and Germany during the eighteenth century, a warning must be given that these movements left the average beekeeper in those countries as little affected as were the peasants of England. The use of wicker hives, log hives, skeps and timber hives continued, and the improved hives were found only in the apiaries of the learned class who could have and read the new books. (Fraser 1950)

A Swarm of New Magazines

One of the ways that science and practice were to join was in the emergence of periodicals dedicated to the cause. *The American Farmer* began publication in 1819, in Baltimore, MD. Initially, these weekly papers relied heavily on copy, plagiarized without attribution. The writing of British beekeeper Robert Huish was being reprinted in the new publication. Chief among his messages was the brutality of the average beekeeper, who knew of no other way to get honey but to kill the bees. He regarded the beekeeper of the day thusly:

They cannot be supposed to possess an intuitive knowledge of the niceties of the science, and all the skill which they do possess, appears to have been inherited from their grandmothers and great grandmothers, who wisely concluded, that if they placed their hives in a garden—they had nothing more to do than to watch their swarms, and then to suffocate them—and should any disaster befall their hives—their want of management and skill was the last thing which entered into their heads; their son or daughter had seen a witch riding through the air on a broomstick, and the Bees had certainly been killed by the indignant Hecate. (Huish 1820)

Robert Huish was a skep beekeeper, who wrote for other beekeepers. However, according to Fraser, his practical advice made him a trusted ally of the common beekeepers who were still skeptical of the scientists, or philosophers as they were called, whose discoveries under the microscope seemed fantastic and not necessarily believable.

This is one of the reasons it took centuries to demolish the notions of the King Bee and the ignorance regarding honey bee sexuality. Even after it was determined that the queen is female and the drone is male, people were unable to discover when or where copulation took place. It took the diligence of the blind naturalist François Huber and his ably sighted assistant to carry out experiments, published in French in 1806. They proved that the queen bee mates high in the air, at the beginning of her life and does not go on mating flights again after she begins laying. What is especially interesting is that so many of these discoveries were made before the invention of the modern hive, which is designed to facilitate close observation.

Works Cited

Boynton, Henry Walcott. (1931). Annals of American Bookselling 1638– 1850. New York John Wiley & Sons, Inc.

Butler, Charles. (1609). The Feminine Monarchie. Or a Treatise concerning bees, and the due ordering of them. Oxford. Printed by Joseph Barnes.

Croker, T. H., Williams, T., & Clark, S. (1766). The complete dictionary of arts and sciences.

Fraser, H. M. (1958). History of Beekeeping in Britain. London, Bee Research Association, Ltd.

Horn, Tammy. (2005). Bees in America: How the Honey Bee Shaped a Nation. Lexington, Ky. : University Press of Kentucky.

Huber, François. (1806). New Observations on the Natural History of Bees. Edinburgh : Printed for J. Anderson ; London : Longman, Hurst, Rees, and Orme.

Huish, R. (1820). The cottager's manual for the management of his bees: for every month in the year; both on the suffocating and depriving system. Wetton and Jarvis.

Knight, David K. (1989). Natural science books in English: 1600-1900. London: Portman.

Lawson, William. (1617). A New Orchard and Garden; with The country Housewife's Garden. London: printed by B. Alsop for R. Jackson.

Mason, P. A. (1998). American Bee Books: An Annotated Bibliography of Books on Bees and Beekeeping From 1492 to 1992. Cornell University Dissertation.

Minor, T[haddeus]. (1804). The Experienced Bee-Keeper; or, a Short Treatise on the Management of Bees [&c]. Litchfield [Conn.]: Printed at T. Collier's Office. 21 p. Pamphlet.

New-Hampshire Spy. (1787). Portsmouth, N.H. : George Jerry Osborne.

Pinkett, H. T. (1950). The" American Farmer," a Pioneer Agricultural Journal, 1819-1834. Agricultural History, 24(3), 146-151. Chicago

Pryor, Elizabeth Brown. (1983). Honey, Maple Sugar and Other Farm Produced Sweetners [sic] in the Colonial Chesapeake. Accokeek Foundation.

Skinner, John Stuart. (1819). The American farmer; devoted to agriculture, horticulture and rural life. Baltimore; Washington, S. Sands, etc.

Wright, Thomas Goddard. (1920). Literary Culture in Early New England 1620-1730. New Haven: Yale University Press.

Chapter 3

The Early Years of Beekeeping in the United States

The 1800s - A New Century

Despite the steady pace of discovery, there is no evidence that general beekeeping practices were evolving to any great extent, especially not in the United States. Many beekeepers still kept bees in single skeps, boxes, or log hives. Swarming was regarded as the chief or only way to obtain new colonies. Many people seemed oblivious to the methods of progressive beekeeping, which involved managing the hive space to increase the capacity of the hive as needed, and which eschewed the killing of the bees to obtain the honey harvest. Undoubtedly, busy farmers felt that the old ways worked well enough, and if one wanted more honey, one simply acquired more bees - not newer techniques. Be that as it may, there were some who questioned the old ways.

An early pioneer in rural Pennsylvania, David Souder was also an avid beekeeper. He took on the task of translating the work of a German author, the Rev. J. L. Christ, and published a booklet in 1807 which he called *The Rural Economist's Assistant in the Management of Bees*. This is one of several publications which I found available only in the form of microprint, where very tiny images of the pages have been printed on photographic paper. Some type of magnifying scanner is required to view and print the pages. Some of the publications are undoubtedly too fragile to view any other way, without risk of them disintegrating. Thankfully, people had the foresight to preserve them in this way. The author begins with this lament:

> I was often led to wish, that their honey might be obtained, without destroying the creature which gathered it; not only because it appeared to me, as too great an outrage committed on the animate creation, to destroy a whole nation of the most inoffensive and industrious creatures upon earth, for the sake of a small booty; but I thought, that the creature, by being let live, might repay our generosity, or rather our doing it justice, by its future usefulness. (Souder 1807)

We have no way of knowing how widely read this slim volume was, but it appears that it was influential. By 1828, it was said: "In the immediate vicinity of Philadelphia, the sectional boxes of Christ are employed almost exclusively. The tract of Souder is a very useful practical treatise, written in a style of winning simplicity, and indicative of an amiable disposition. It is the manual of the bee-masters of this state" (*American Quarterly Review* 1828).

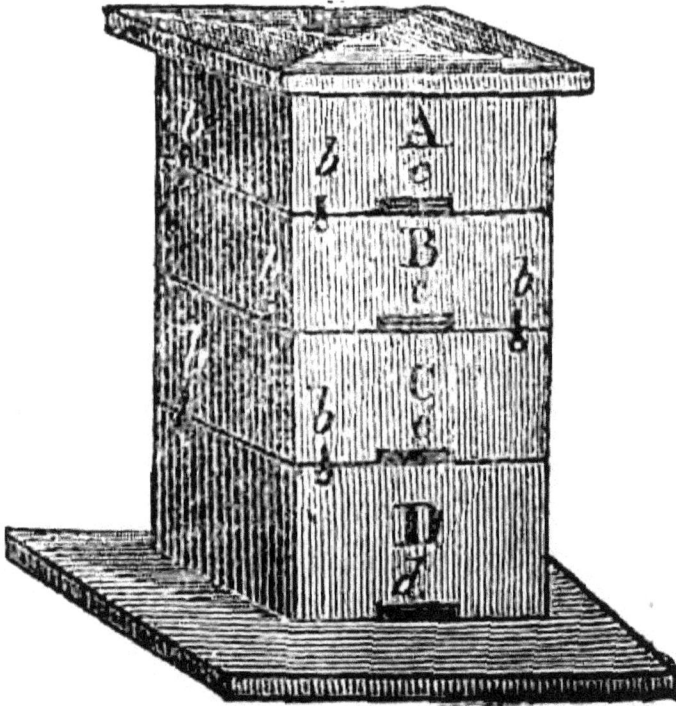

Christ's tiered Hive

Dr. Joseph Doddridge, living in the "far west" of Ohio in the 1800s, wrote a small pamphlet called "A treatise on the Culture of Bees" (1814). He recounts:

> From my infancy I was fond of the bees, and accordingly, as soon as I became a house-keeper, I procured a stock of them, and for a while managed them in the usual way. I was however, always shocked at their destruction for the sake of their treasure. This is too hard a fate; said I to myself. Is there no way of avoiding it? Cannot I take a part of their labour, and leave them enough behind to support them through winter? Must the acquisition of their treasure on my part be preceded by the death of the little labourers? I hope not.
> (Doddridge 1814)

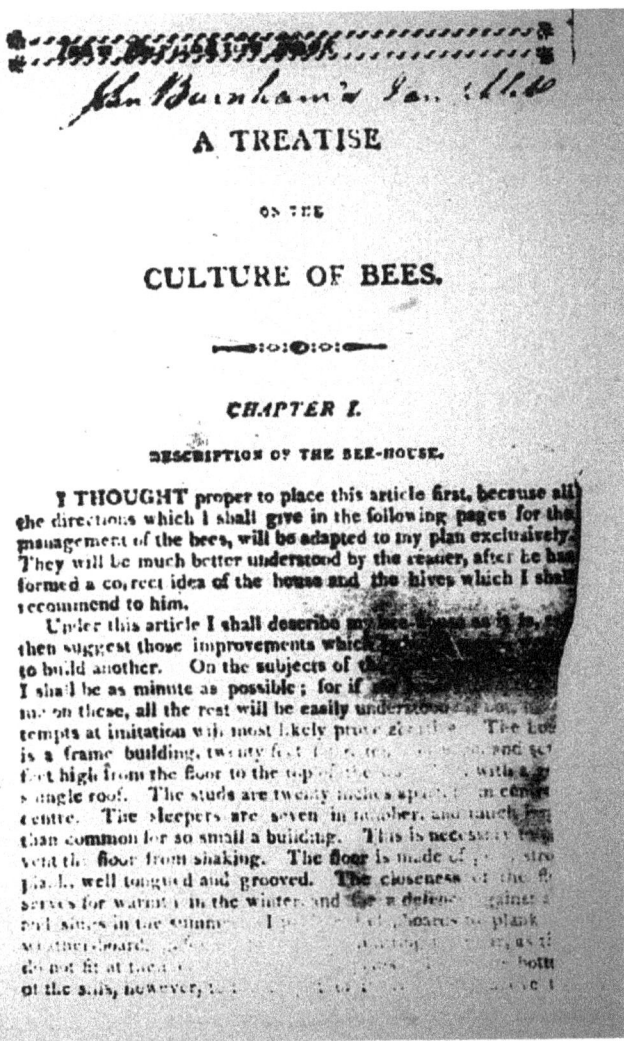

Doddridge treatise title page

The author begins his work with the description of an elaborate "bee house." Unfortunately, the copy I have is illegible in many sections, but much can be gleaned. After constructing a building for his hives, he paints it overall with what he refers to as "Spanish brown." But then, he describes the openings in the wall through which the bees would pass from inside to the outside world. These were adorned with distinctive half circles, which he painted in "all the different colors he could procure, for the purpose of enabling the bees to distinguish their respective hives." In other words, whereas they would have trouble recognizing a simple opening in the monotonous brown wall, he believed these colored half circles would be recognized and remembered as much as different colored houses set them apart from one another. Certainly this is an example of remarkable thoughtfulness on the part of Dr. Doddridge, considering the fact that in general, no attention was paid to the external appearance of bee hives at that time, other than to make them all the same – or simply slap them together with whatever scraps of wood were handy.

He describes the wall of the bee house as having twelve arched entrances, colored in white, black, red, green, blue, orange, and so on. The hives are placed inside the building, presumably on shelves as was commonly done in Europe. Unfortunately, much of the pamphlet is in poor condition and illegible. Would that I could obtain a pristine version!

The American Farmer

> *The appearance in 1819 of a periodical bearing so novel a title*
> *as the American Farmer excited curiosity and suspicion. An aged*
> *gentleman commented, "What won't they get a going next!"*
> *A farmer asked, "What is it?" A bystander replied, "Why a thing*
> *to coax us farmers to vote them into office, I s'pose." Although*
> *the American Farmer was supported by the patronage and pens*
> *of distinguished agriculturists in America and abroad, it failed to*
> *receive the support of the great mass of small farmers. (Pinkett 1950)*

The inaugural issue of *The American Farmer* had a short piece on beekeeping: "A method of taking the honey without destroying the Bees."

> *The common practice of killing the bees, in order to obtain the*
> *honey, few can witness without some little compunction; and as*
> *there is a very simple method of effecting the object, without any*
> *injury to this most interesting little animal. (Skinner 1819)*

The author describes his technique of blowing clouds of smoke from a lit cigar deep into the hive, which has the effect of stupefying the bees into submission, without actually killing them. The beekeeper could then judiciously remove large pieces of honey comb, leaving the bees to replace the comb in the cavity left by its removal. The author claims "The above method has frequently been practised by myself and others, and have always found it to do well." He signs the piece with a pen name: AMATOR MELLIS, which is sort of Latin for "lover of honey."

The *American Farmer* was to lead the movement to advance agriculture. One of the ways was to publish specific queries from readers in order that other readers might provide appropriate responses. Appearing in the September 17, 1819 issue was this entreaty:

> SIR,—In this section of the country our bee-hives are infested with a webworm, which, when fully grown is about an inch long, they are in such quantities as to take entire possession, drive off the bees, destroy all the honey and finally the wax. Will you be so good as to give a hint in your paper and request any of your readers, that may be possessed of a remedy, to publish it in the American Farmer. (Skinner 1819)

The forthcoming reply the following year in the May 19 issue was sincere, if unhelpful;

> To prevent Bees being destroyed by the worm. Set each corner of the hive on a piece of wood about an inch square and a quarter of an inch in thickness,-- this elevates the hive a little and prevents the deposits of the eggs which produce the worm, and which are always placed where the edges of the hive come in contact with the bench. (Skinner 1820)

While it is true that moths lay their eggs in crevices, there are plenty of other crevices in a hive besides the one our author singles out as the culprit. In any case, wax worms are usually a symptom of lack of vigilance on the part of the bees. Populous healthy colonies defend themselves against these invaders where small, dispirited ones may not. But this did not prevent entrepreneurs from inventing and patenting hundreds of "moth proof" hives, none of which addressed the real problem, and most of which were overly complicated and even exacerbated the problem when bees failed to flourish in these contraptions.

Continuing, in June there appears another letter about taking honey without killing bees, a recurring theme. While the previous correspondent seems to be talking about bees in a single box, this one describes a sectional hive consisting of three compartments stacked up. The writer, referring to himself as "A Subscriber," pries the topmost box slightly apart from that below and pulls a fine wire all the way through, neatly cutting the combs between the boxes, making it possible to lift the honey-filled box off of the hive. He then allows the bees to fly back to their proper hive, leaving him with the harvest.

Finally, he cautions us to "leave the bees enough honey to carry them through the winter." This is pretty clear evidence that by 1820, beekeepers were building sectional hives out of wood, that they understood that the honey tended to accumulate in the upper section, and it could be removed without ruining the bees' prospect for survival. Many additional refinements were to follow, each aimed at making the process less destructive to the bees and the combs they worked so diligently upon but which were too attractive to the beekeeper to pass up. But this gives a glimpse of the incremental progress taking place at the beginning of the 19th century.

Huber's leaf Hive

Huish and Huber in America

On March 29, 1822, the *American Farmer* began serializing Robert Huish's *The Cottager's Manual, for the Management of Bees Throughout Every Month of the Year*. Despite being written by an Englishman for a specific audience, much of the advice it contained was universal and still applicable today:

> Before I enter upon the particulars of the practical management of Bees, it may be attended with great advantage to the cottager, to receive some instructions which ought to be minutely attended to in the purchase of hives; for one of the first things which it is necessary for a person to know, who is desirous of keeping Bees, is the difference between a good and a bad hive. There is no kind of stock in which a purchaser can be so easily deceived as in a hive of Bees, for few possess either the courage or the ability to examine it, and still fewer are aware of the particular excellencies which ought to distinguish a sound and healthy hive. (Skinner 1822)

The entire work was made available to the American reading public, at minimal cost, through the medium of a weekly newspaper. Subsequently, they brought to notice the work of Francois Huber, which had been translated from French to English and published in 1806. There occurred a brief mention of his discoveries embedded in a reprinting of some "Letters from Paris, April 1822." Appended was the comment: "The following letter contains so many curious facts in natural history, &c. that we beg to request attention to its compressed intelligence."

The letter ranges from such topics as the electro-magnetic experiments of Ampere, the discovery of new comets by astronomical observer Pous, and the work of Huber, duly noting that Huber was blind but ably assisted by his family. They declaim: "The Hubers have been for these twenty or thirty years, in the habit of making observations on bees, and it is probable that they will in due time succeed in discovering all the secrets of the hive."

Among his revelations were included the fact that honey bees do not gather beeswax from flowers as was commonly supposed, but that the wax is produced via tiny glands in the abdomen. Further, bees do collect pollen, referred to by writers of the time as farina, and it is required to make the food for the developing larvae, which are unable to feed themselves. Writing in a subsequent issue, a reader thanks the journal for including the extensive writing about bees. Further, he also mentions Huber:

> *An author, who, notwithstanding the sneers and criticisms against him, I will venture to say, has thrown more light on this particular subject of natural history, than any one before or since his time. Several subsequent writers have presumed to question, and sometimes to contradict his assertions, for want of that peculiar skill, address and perseverance which belonged to Huber and his assistant. But any gentleman who hath both leisure and inclination to acquaint himself with the admirable order and economy of Nature in this emblem of industry, should consult Huber, and may safely take the results of his experiments as his Code. (Skinner 1822)*

This last statement makes it plain that the editors are seriously attempting to distinguish between fact and fantasy, and advise the readers where the better quality information can be obtained.

Home Grown Beekeeping

American bee books also began to appear in the 1800s. The following two reviews cover publications from the Boston area:

A Practical Treatise on the Management of Bees; and the Establishment of Apiaries, by James Thacher.

> *Much has been said of the fittest size and form of a hive and of the materials of which it should consist. Those composed of wood, straw, husks, and osier, have been at various times recommended; but at present, sectional boxes placed in tiers, one above another, or hives provided with boxes on the tops, are generally considered preferable. (Thacher 1829)*

The inventive beekeepers of the time were making modifications to their box hives, which foreshadows the future of beekeeping. One of the problems with traditional hives is that the bee combs, being made of soft wax, are susceptible to breakage and collapse, especially when filled with heavy honey. The basket-like skep hives were often equipped with cross sticks which lent some support to the combs. The enterprising beekeepers that used wooden box hives began to place wooden slats across the tops of the boxes. These slats were cut about the width of natural combs (1 inch) and separated from each other by a space reflecting the gap that honey bees place between their combs. This gap is about the size of two bees. This enables them to work back-to-back on the opposing comb faces.

Above these slats, honey producers placed small boxes which the bees could fill with honey, making a nice neat saleable package when removed from the hive and emptied of bees. Unfortunately, honeybees are great improvisers and often do not work according to their master's plans. They might attach their combs to the bars in a neat orderly way, or crisscross the bars at angles, or deviate in curvy convolutions. Usually the queen bee would stay below the bars in the "brood nest," and the bees would fill the boxes with marketable honey. Sometimes the queen would wander about the hive looking for empty comb in which to lay eggs. This leads to larvae and pupae mixed in with the honey, making it unsaleable as is. The combs could still be squeezed and the honey strained through wicker to purify it. An interesting observation made by Dr. Thacher reveals that American beekeeping was still in its infancy:

> Even in America, honey and wax are imported to a very considerable amount, although no country possesses greater advantages for the culture of the bee, and by proper attention to the subject, the necessity for importation might be entirely superseded.

> There is no doubt that bees were formerly more frequently kept in America, than at present. In many places in New Jersey, where there is now scarcely a bee to be seen, there once existed millions of these insects, to the great profit of their owners.

> It was common for one dealer in a country town to sell from fifteen or twenty barrels of strained honey alone, to say nothing of wax and comb honey, as well as a kind of wine, made of the washings of combs, called metheglin. These articles of commerce have almost disappeared, and we find that it is mainly attributable to the ravages of the miller, or night moths, which have of late years spread destruction through the hives. (Thacher 1829)

He states that bees were widespread formerly, though this may be an exaggeration. Much honey and wax was certainly harvested from wild honey bees living in the woods, but the craft of beekeeping was not widespread nor was it as sophisticated as the work of Thacher would imply.

An Essay on the Practicability Of Cultivating The Honey Bee, In Maritime Towns And Cities, As A Source Of Domestic Economy And Profit by Jerome V. C. Smith.

This inventive beekeeper gives explicit directions for the construction of glass hives "for the purpose of inspection, ... studying their habits, their labors and their government." He suggests placing over an ordinary wooden box hive, a glass globe which would hold ten or fifteen quarts (nearly four gallons!) but he claims to be in possession of an even larger one holding a bushel. How big this would be we can only guess, but today a bushel is reckoned to be equivalent to 9 gallons, which would be a very big glass vessel. A quick look online tells me that 7-gallon wide mouth glass carboys are readily available, though I would not be interested in trying this myself, as such a container filled with honey would likely weigh 60 pounds or more.

It is a fascinating book, which has much to offer the reader. He promises to tell us of their government, but in the end is puzzled by it.

> *Notwithstanding an extraordinary attention to the construction of the glass, which magnifies the bees considerably where it is most convex, I never have discovered the least clue to the mode of government. (Smith 1831)*

But of course, that does not prevent him from speculating on the question!

> *I think the government generally, and certainly all special commands, are first made and propagated by the appropriate officers, by striking the horny tip of the tail on the hive or comb —so that a tremor, differently modified, gives a general as well as instantaneous information, which every bee not only perfectly understands, but quietly obeys. (Smith 1831)*

According to the Rev. Lorenzo Langstroth, whose work will figure prominently later, this was the first book on bees that he purchased, in 1835.

Works Cited

Doddridge, J. (1813). A Treatise on the Culture of Bees. Ohio. [Early American imprints. Second series ; no. 28342].

Huber, François. (1806). New Observations on the Natural History of Bees. Edinburgh : Printed for J. Anderson ; London : Longman, Hurst, Rees, and Orme.

Huish, Robert. (1820). The Cottager's Manual, for the Management of Bees Throughout Every Month of the Year.

Pinkett, H. T. (1950). The" American Farmer," a Pioneer Agricultural Journal, 1819-1834. Agricultural History, 24(3), 146-151.

Skinner, John Stuart. (1819-1834). The American Farmer; devoted to agriculture, horticulture and rural life. Baltimore; Washington, S. Sands, etc.

Smith, Jerome V. C. (1831). An Essay on the Practicability of Cultivating the Honey Bee, in Maritime Towns and Cities, as a Source of Domestic Economy and Profit 1st. Boston: Perkins and Marvin; New York: J. Leavitt.

Souder, David. (1807). The Rural Economist's Assistant in the Management of Bees, principally taken from the German writings of the Rev. J. L. Christ. Lancaster, [Pennsylvania]: Printed by William Greear [Early American imprints. Second series; no. 13618].

Thacher, James. (1829). A Practical Treatise on the Management of Bees; and the Establishment of Apiaries, With the Best Methods of Destroying and Preventing the Depredations of the Bee Moth. 1st. Boston: Marsh & Capen.

Chapter 4

The Coming Revolution

Before Langstroth

Most beekeepers have heard the name Langstroth. L. L. Langstroth was born in 1810 in Philadelphia, and died in Dayton, Ohio in 1895. On October 20, 1951, a memorial bench was dedicated to him with the following inscription:

> In bee hives of a century ago, bees fastened comb to the hive walls. Removal was possible only by cutting, often after killing the bees. Management of colonies under this crude procedure was difficult, and apiculture languished. In 1851 Langstroth discovered a new basic fact of bee behavior: They respect and leave open any hive spaces 3/8 inch wide, whereas they seal narrower ones with bee glue and utilize wider ones for comb. Brilliantly applying this fact, Langstroth designed the moveable-frame hive, the frames separated from each other and from the hive walls by the inviolate "bee space." With its comb-containing frames now freely move-able without injury to bees or comb, the Langstroth hive ushered in a new era in bee culture. (Langstroth Memorial Bench, 1951)

While it is true that prior to Langstroth's invention, beekeeping was still primitive, it was not languishing. Books were coming out regularly describing the biology of bees and the techniques of producing honey from so called box hives, the insides of which were inaccessible except by cutting. This contrasts with a beehive today, which appears to be made up of just boxes but actually contains bee combs mounted in wooden frames in order that they may be examined and removed as appropriate. While Langstroth's discovery did launch the beekeeping revolution, this was entirely because of the fertile environment in which it appeared.

I will describe three contemporaries of Langstroth; each advanced bee culture of the time, around 1850. Of these three, only Moses Quinby is generally remembered for his contributions. In my view, T. B. Miner's part in the story is large but principally because of his role as a writer and publisher. The third, Kelsey, is only included to give a sense of what the times were like when this story takes place.

William R. Kelsey

I basically stumbled upon the work of William R. Kelsey, of Syracuse, NY. His work is titled *The Apiarian's Guide, being a practical treatise on the culture and management of bees*. Kelsey leads with a quote from the Report of H. L. Ellsworth, Commissioner of Patents:

> *No branch, perhaps, of agricultural or rather rural occupation, has been so much neglected in this country as bee culture. Wherever it has been attempted with care it has always proved profitable; but many who engage in this business, abandon it. (Kelsey 1847)*

Kelsey gets right to the point when he says "some fifty or sixty volumes have been written upon the subject of bees" and "there are almost as many opinions among the mass of bee culturists … as there has been authors upon the subject." How little has changed; beekeepers tend to be an opinionated and vocal group even today. He goes on to point out that so much that has been written on the topic is "so mixed up with error … as to be of little or no benefit." He depicts the readers as consumers of information with little discernment nor regard for facts. Then, he proceeds to commit a huge blunder.

Kelsey describes the prevailing theory that the worker bees can take an ordinary bee egg, destined to be just another worker, and by means of the shape of the cell in which it's raised and the type of food the larva is fed, it miraculously is transformed into a queen bee. This, of course, is true; whomever was promulgating such a story was correct. But this is simply unbelievable to Kelsey who declaims it "contrary to every idea of reason."

> *As well might we believe in the possibility and probability of converting a mule into a breeding mare merely by giving one a large stall and "more pungent food"!! (Kelsey 1847)*

In short, his treatise resembles closely the books he describes as mixed up with error and lacking in benefit. It is riddled with so many other errors as to render it a comical book, to be read with curiosity perhaps, and a tinge of sadness that he could have been so adamant and so wrong. Another epic folly is his projection of how honey beekeeping might be expanded by simply doubling one's holdings each year. Commencing in 1847, he supposes that one might have 2,048 hives by 1857. He assures his reader that:

> *Two swarms of bees, with two of my hives, and an individual right,*
> *will cost about $15, which within about ten years will realize to a man*
> *a handsome fortune over and above all expenses … the apiarian*
> *would be worth $16,384 over and above all loss and expenses of*
> *every kind. (Kelsey 1847)*

Not content to let the facts and numbers speak their own story, he graces his slim 46 page book with about ten pages of testimonials, principally from his neighbors in the vicinity of Dundee, NY. Among these he includes words of praise from Martin Holmes, the Sherriff of Yates County, NY; Reverend Philander Shedd (formerly of Truxton, Cortland Co. NY and more recently of Tompkins Co.), a talented and distinguished Baptist minister, at present located at Dundee, NY; and the Reverend Pastor of the Baptist Church, Burdett, NY. And lastly, a notice from the Dundee Record of June 1846, copied and cordially endorsed by the Syracuse Star, the Penn-Yan Democrat, the Yates County Whig, and "many other newspapers in NY." Kelsey evidently read the same books as the beekeepers of his era, mentioning the English author Edward Bevan, whom Langstroth also credits as an influence. But not without sniping, stating that Bevan's book is:

> *So lumbered up with the ideas, writings and quotations of others,*
> *and contains so many opinions of distinguished "scientific" men, as*
> *to be of comparatively little use for practical purposes. (Kelsey 1847)*

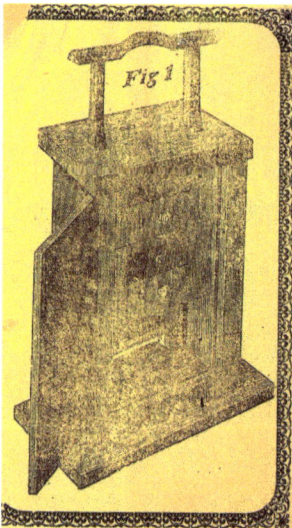

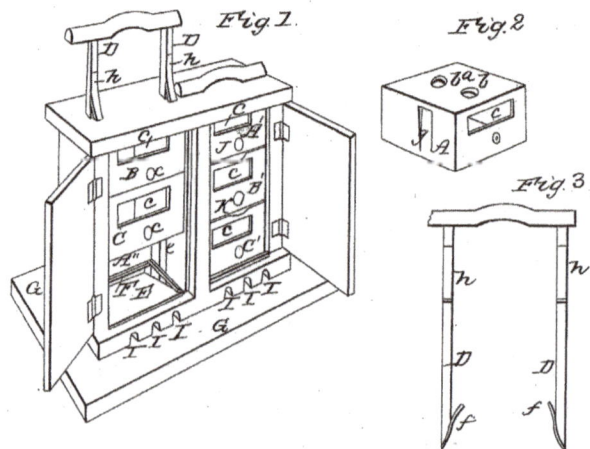

Kelsey Hive and Patent

If I have expended an inordinate amount of time on Mr. Kelsey, perhaps because he and his devotees are from the region in and around Dundee, NY. My grandparents owned a summer house there, on Keuka Lake, about an hour's drive from where I now live.

Thomas B. Miner

At one time, New York City was a rural village, called New Amsterdam. The streets were mud and there were pigs, dogs and chickens everywhere. In 1800, the city had a population of about 60,000 but in the ensuing 50 years it exploded to half a million people. Still, right across the river on Long Island, there were farms and immense nurseries which were importing, growing and selling plants from all over the world. Thomas Miner worked in New York City, but lived in the neighborhood of Ravenswood, which is now part of Astoria in Queens. He was evidently an avid beekeeper and began to write for the American Agriculturist in 1846. He started out a modest exposition on the management of bees, but it continued for nearly 20 issues:

> The art of managing bees in this country is but very imperfectly understood, so far as profit, health, and productiveness are concerned. It is generally supposed that bees require little or no air, and if they prove unproductive, or are lost from the ravages of the bee-moth, it is a mere matter of chance, wholly beyond the control of the owner. I now propose giving the result of my own personal experience in the management of bees for some years, on Long Island; and from the happy effects of my course of procedure, I think my remarks will not prove wholly void of interest, or advantage, to those who are unsuccessful. (Miner 1846)

Mr. Miner never really lets on how many hives he maintains but it is clear that he learned everything he possibly could about them; both by reading and by practice. He kept them in his spare time, when he was not in Manhattan. Miner was born in 1808 and began keeping bees in Long Island in the 1830s while working near Wall Street, in New York City. He also was keen on the culture of poultry, and had the sense that beekeeping and chickens were ideal for small holders seeking to increase their income on a farm or semi-rural lot. In the course of his writing for the *American Agriculturist*, he conceived the idea of writing a structured manual. In order to ensure its completeness, he undertook to visit beekeepers.

On a recent tour through the State of New York, I made it a point to call on every bee-keeper in my route, that I could visit conveniently, merely to gratify a curiosity that I felt, to see how they generally managed bees. I elicited their management by simple questions, and they generally took great pains to give me all the information in their power; for I never ventured to play the teacher, but humbly and civilly received instruction from them, such as they were able to impart, being a stereotype of the management that was in vogue centuries ago, to a great extent. (Miner 1849)

By this he means, bee culture had not advanced significantly during these centuries and Miner set about to remedy this situation. The fact that he lived in and wrote about New York is key to why it would be the epicenter of the beekeeping revolution which led to the Golden Age of Beekeeping.

DEVOTED TO AGRICULTURE, HORTICULTURE, FLORICULTURE, BEES, POULTRY, &C.

T. B. MINER, Editor, Clinton, Oneida Co., N.Y.   UTICA, N. Y., JULY, 1853,   VOLUME 11 NO. 7.

Miner Northern Farmer

T. B. Miner's *American Bee Keeper's Manual* (1849) had a widespread influence and justly so. Not only was it comprehensive and well organized, it had detailed information on the relationship between the environment and the success of the honey bees and their owners.

I would observe, that in different parts of the country, the labors of bees vary according to the bee-pasturage about them. In a location where the white clover (Trifolium repens) abounds profusely, as in Herkimer county, State of New York, and some other great grazing counties, a swarm will produce much more honey and wax, than on Long Island, where the honey harvest is not so abundant.

Of all the resources of bees, nothing can equal the white, or Dutch clover, that abounds to a greater or less extent, throughout the whole country; I may almost say, that without the existence of this flower, it would be useless to attempt to establish an apiary; yet there is no section of the country where it does not exist; consequently, there is nothing to fear on that point. In any place where this clover is found growing in spontaneous profusion, there will bees thrive beyond a doubt.

Among the forest resources of the bee in this country, the most conspicuous are the basswood and maple. From the basswood in particular, a great supply of honey is obtained ; and where this tree abounds, in connection with a profusion of white clover, there is the apiarian's true El Dorado. (Miner 1849)

APPENDIX.

MINER'S PATENT EQUILATERAL BEE-HIVE.

Miner's Hive

His use of the term "El Dorado" alludes to the California Gold Rush taking place at the time he was writing. In fact, the name was used in 1850 for one of the counties in the new state. His idea, of course, was that people did not need to seek lost cities of gold in the distant west, when their treasure lay at their feet. But keeping bees was still a small scale undertaking at the time that Miner offered his book. His career path took him from writing to becoming a publisher, beginning in 1852 with his launch of the *American Farmer* based in Utica, NY in the heart of the Mohawk Valley, not far from the aforementioned Herkimer. In the first issue, he announced that the new paper would "advocate the interests of Poultry and Bee-Keepers." Thomas Miner eventually moved back to the New York City area; he died in Linden, New Jersey in 1878.

Moses Quinby

At the time that Miner was advancing the craft of beekeeping, his peer Moses Quinby was living in Greene County, NY, in the village of Coxsackie which is on the Hudson River about 25 miles south of Albany. He had been born in 1810, in Chappaqua, some 30 miles north of New York City and his family moved up the river around 1820. Moses worked in a sawmill, operating a wood lathe and other equipment. His interest in wood extended to making fine furniture from boards of maple and cherry. It was during this period that he commenced beekeeping, and soon was contributing to contemporary periodicals. He got off to an auspicious start when a long article by him appeared in the monthly journal, *The Cultivator*. He began:

> *A knowledge of facts constitutes science. Correct observation alone, can lead to a knowledge of any science; from such knowledge only, will correct practice result. The honey bee has been a prolific theme for guessing among ancients as well as moderns. To refute some of these incorrect theories by a relation of facts, is the object of this communication. (Quinby 1847)*

At this point, he tells us that he has been a beekeeper for twenty years and has owned more than one hundred hives for seven years, so already we can see he has far outpaced his peers. He discounts a number of prevalent errors, including the supposition that pollen (bee bread) is converted to honey. It is not he says; pollen is "food for their young." That is, the food fed to the larvae consists of digested pollen and nectar, both from flowers. It is apparent that he follows the discussions in print and he responds to them directly. He had written to *The Cultivator* previously when in 1842 he addressed the topic of "caterpillars in apple trees."

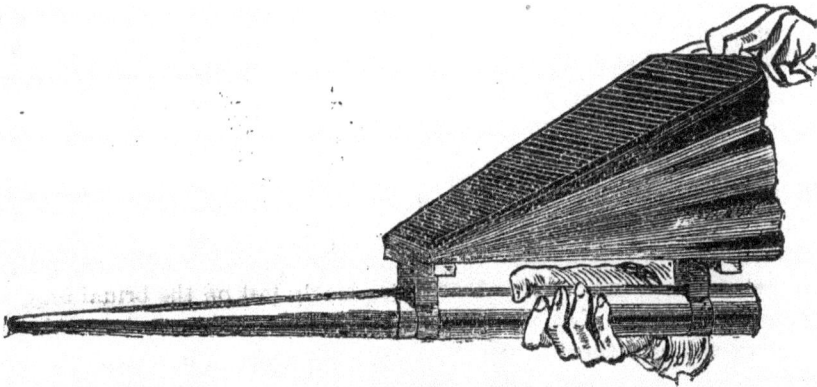

THE QUINBY SMOKER.

Quinby smoker

Mr. Quinby and his family moved to St. Johnsville NY in the 1850s and lived near there till his death in 1875. Like Mr. Miner, Quinby had been drawn to the Mohawk Valley by the prospect of large honey crops. However, he was of quite a different disposition than Mr. Miner. Where the former was principally a purveyor of information in the media of the magazines, Mr. Quinby was a honey producer. His industriousness led him to building hundreds of hives, populating them, and marketing quantities of honey hitherto unheard of. After he died, he was praised for decades, as in this example published 25 years after his death:

> Quinby was one of the most successful bee-keepers who ever lived; indeed, if I am right, Mr. Quinby was the first to produce and ship a whole boatload down the Hudson to New York City. Such an amount of honey at that time (the early 50's) literally broke "the market," for no one knew what to do with so much honey. This remarkable feat was performed with box hives. Yes, father Quinby made his bees pay; and, as I understand from Capt. Hetherington and others who knew him, he was one of the most lovable and unselfish persons the world has known. Never a beginner went to him for instructions without receiving generous advice, even though that advice might bring into his territory new and disagreeable competitors. (Root 1901)

In the foregoing, Quinby is credited with a "remarkable feat" – that of producing such a large crop using the conventional box hive, the use of which had been thoroughly described by Mr. Miner, but had never been taken to the next level of productivity. By 1853:

45

He had so satisfactorily established a system of bee-keeping that would insure reasonable return for a stated investment, that he felt warranted in publishing the first edition of this work entitled "Mysteries of Bee-keeping Explained." Simultaneously with this publication, appeared the first edition of "Langstroth on the Hive and Honey Bee." These two works were the first of any great value that had been written in America. (Root 1879)

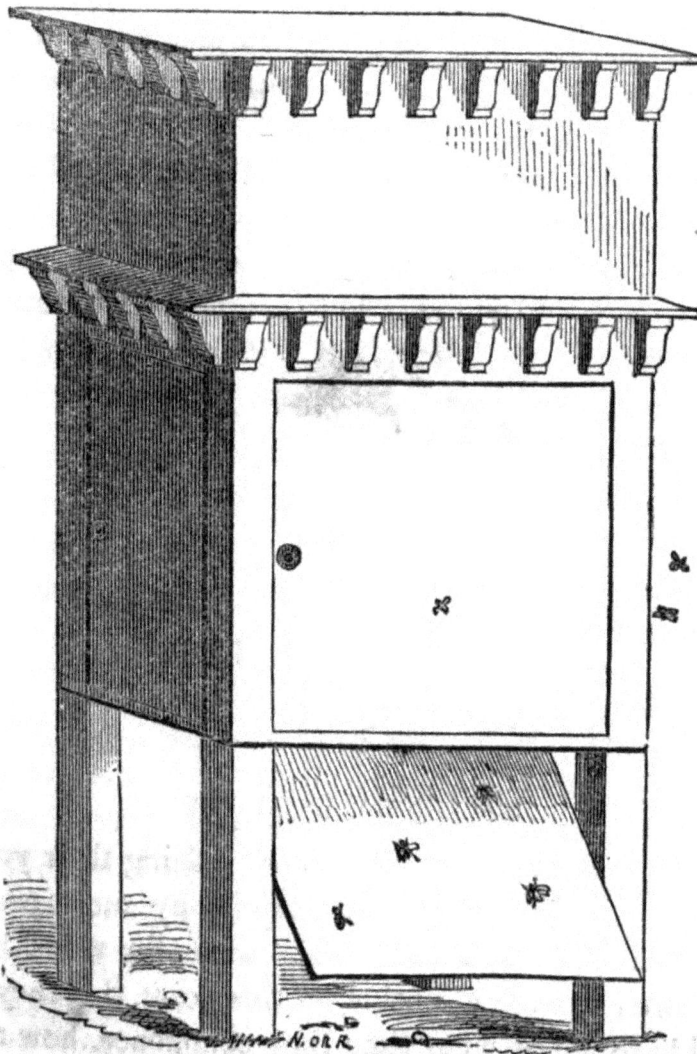

Quinby Hive

In his own words, Mr. Quinby explained:

> *We found that boxes made of glass, in a fanciful style, commanded*
> *a still more ready sale. This fact I had nearly all to myself, and had*
> *I been as shrewd a money-getter as Astor, Stewart, or Vanderbilt,*
> *I might at least have secured a moderate fortune. Instead I wrote*
> *a book of instructions, which I hope has been of some use. This*
> *was called a mistake, by some of my best friends. The middle men*
> *who distributed to consumers, called me a fool for doing so. "Don't*
> *you see that competition will reduce the price, and you will not get*
> *remuneration for what you have done?" (Quinby 1873)*

Mr. Quinby went on to make many contributions to the beekeeping revolution, which will be discussed at length later, but by the 1850s, he had clearly made a name for himself as the frontrunner in the race to modernize beekeeping, which had been carried more or less unchanged for centuries. What the beekeeper of today may find astonishing is the fact that he and his peers used what essentially was a "black box," the internal workings of which were just barely understood.

A keeper of bees of that era knew how to obtain honey by adding space for its storage by the bees. Hives were generally stocked by capturing the swarms of bees that proliferated in the spring, about the time of the apple blossoming. The error of killing bees to take their honey was proven by the success of those beekeepers who learned to leave enough honey with the bees, so that they might survive over winter – and thus, to prosper for many years. Beekeeping lost its reputation as a risky gamble, and people such as Moses Quinby showed by example that a satisfactory living could be earned by keeping bees.

Works Cited

Anonymous. (1851). The hive and its wonders. Written for the American Sunday-School Union. Philadelphia, PA.

Kelsey, W. R. (1847). The apiarian's guide, being a practical treatise on the culture and management of bees. Syracuse, NY. Kinney, Marsh & Barns.

Langstroth, L. L. (1853). Langstroth on the Hive and the Honeybee, a Beekeeper's Manual. Northampton, MA. Hopkins, Bridgman.

Miner, T. B. (1846). The American Agriculturist. Vol 5, p. 213. New York, Saxton and Miles.

Miner, T. B. (1849). The American Bee Keeper's Manual: Being a Practical Treatise on the History and Domestic Economy of the Honey-bee. London. John Wiley.

Naile, F. (1942). The life of Langstroth. Ithaca, NY: Cornell University Press.

Newman, T. G. (1886). Bee literature. *American Bee Journal*. 22, 663. Chicago, IL.

Quinby, M. (1847). False theories in relation to bees. The Cultivator. Vol 4. Albany, NY. Luther Tucker.

Quinby, M. (1853). Mysteries of Bee-keeping Explained: Being a Complete Analysis of the Whole Subject. New York. C.M. Saxton.

Quinby, M. (1873). President Quinby's Address to the North-Eastern Bee Keepers' Association. *American Bee Journal*. Vol. 8, p. 202.

Root, E. R. (1901). Gleanings in Bee Culture. Vol. 29, p. 909. Medina, OH.

Root, L. C. (1879). Quinby's New Bee-keeping: The Mysteries of Bee-keeping Explained. Orange Judd Company.

Chapter 5

The Beekeeping Revolution

Langstroth and His Hive

Langstroth

Beekeepers the world over use a version of the Langstroth Hive. Yet, few of them know of the life and times of the Reverend Lorenzo Langstroth (1810-1895). An excellent book, *The Life of Langstroth* was published to very little acclaim in 1942, during the World War. It was subsequently reprinted in 1976, by Cornell University, which is when I acquired my copy. For those wishing a thorough acquaintance with the man, there is no better place to look. In the introduction of her book Florence Naile began:

> Before the days of Langstroth there were many books, now chiefly interesting as history or as curios. Moses Quinby issued his noteworthy book, *Mysteries of Beekeeping*, in the same month in which Langstroth published the first edition of his classic work [Hive and the Honey-Bee]. Every movable-frame hive in the world tells mutely the story of his genius, yet beekeepers of today know almost nothing of the man himself, who toiled that they might reap.
> (Naile 1942)

Other authors have looked into the history of Langstroth and his hive, and have come away with a deep sense of how ingenious but troubled this man was. Sue Hubbell, in her *Book of Bees*, delved into Langstroth's life, as many have, and came to his handwritten journals. They have recently been scanned and placed online for all to see. Unfortunately, they were hastily written with a fountain pen and are difficult to read for anyone unaccustomed to the style of penmanship. To make matters worse, large sections have faded badly. Sue Hubbell thought of the writings as "the key to an understanding of the man." She proposed bringing forth an edition of his journal, but her editor was adamant that such a book must leave out his profound psychological difficulty, leading her to drop the idea altogether. Sue Hubbell observed:

> But it was this area that fascinated me: how a man so divided against himself could nevertheless contribute more useful knowledge and craft to the world in what was functionally only half a life than the rest of us, who are presumably in mental good health, do in the whole of our ordinary lives. (Hubbell 1998)

Like many brilliant creative people, Langstroth was deeply affected by what he referred to as his "head trouble." It might be diagnosed today as chronic depression, for which there was no reliable treatment. Sometimes it would completely incapacitate Langstroth, leaving him unable to work.

At the age of 26, Langstroth accepted an urgent request to be the pastor of the South Church in Andover, MA. The same year, he married Anne Tucker, and they had a son the following year. According to Naile, he found this new undertaking to be far too difficult. The large congregation comprised 500 people, the obligations were excessive, and his recurring psychic distress led him to depart after less than three years. He found employment tutoring and teaching but had difficulty holding a job due to his changeable mental state. Years later, Langstroth was to recount his introduction to beekeeping at this point in his life:

> In the summer of 1838 the sight of a large glass globe, on the parlor-table of a friend, filled with beautiful honey in the comb, led me to visit his bees, kept in an attic chamber; and In a moment the enthusiasm of my boyish days seemed, like a pent-up fire, to burst into full flame. (Langstroth 1893)

In 1848, the family moved to Philadelphia, where Langstroth was born, and he and his wife opened a school for girls. Langstroth continued his pursuit of beekeeping and adopted the hive in general use in the Philadelphia area, after the fashion of the German Rev. Christ, as described and promoted by David Souder. The hive had a large roof with openings over which Langstroth would place multiple glass jars that the bees would obligingly fill with honey. The hive also had the improvement promoted by Edward Bevan, namely wooden bars the bees could attach their combs to, rather than affixing them to the roof as in earlier hives which were simply plain wood boxes, with flat board roofs.

Florence Naile relates: "Beekeeping was diverting him more and more. The hive held baffling but fascinating problems." In the course of devising a way to keep the bees from gluing the roof to the comb bars with sap they collect from trees, he discovered that if he left a gap of approximately 1/4" to 3/8" (6-9mm) between the bars and the roof, they would keep this crawl space free of comb and glue. This led him to the idea of mounting the bee combs in wood frames which would hang in the hive like hanging files in a file cabinet but separated from the hive sides and each other by this gap, which became known as the "bee space." In his own words, he explains the merits of his invention:

> Movable frames. - To be able to remove the combs from the hive without mutilating them or seriously disturbing the bees, will secure the following advantages in the management of an apiary: The combs may at any time be readily examined for any purpose. Feeble colonies may be strengthened, by transferring to them from stronger colonies, combs containing honey and maturing brood. The queen of a hive may be easily caught, for any purpose. Spare comb and honey may be easily removed. New colonies may be multiplied to an extraordinary extent, the apiarian being made independent of the uncertainties and perplexities of natural Swarming. In short to be able easily to remove the combs, and to transfer them from one hive to another. (Langstroth 1853)

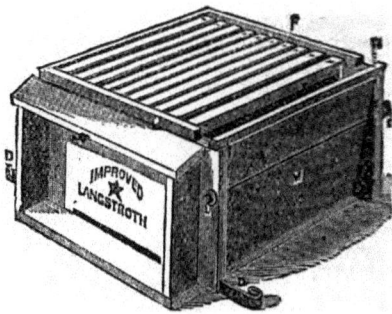

Langstroth Hives

Not only was this new way to access the hive a boon to the beekeeper, it was equally better for the bees. In the past, beekeepers had an antagonistic relationship with the bees, and all the cutting of combs and rough handling tended to either infuriate them or greatly interfere with their successful functioning. A hive with freely moveable combs made the examination of the hive and its contents far less disruptive. A colony living in a modern hive, tended by a conscientious beekeeper, is simply a much better arrangement. Of course, people can still rough handle bees, infuriate them, and generally make their lives miserable. Langstroth lamented:

> *I am perfectly aware how difficult it is to reason with a large class of bee-keepers, some of whom have been so often imposed upon that they have lost all faith in the truth of any statement which may be made, while others stigmatize all knowledge which does not square with their own, as "book- knowledge," and unworthy the attention of practical men. (Langstroth 1853)*

However, not everyone viewed "book-knowledge" with such a dim view. Langstroth's book was first published in 1853 and revised many times over the ensuing decades. It was enthusiastically received by the segment of the public whose appetite for learning was insatiable.

> *The Langstroth hive is the best, and his book on the honey bee is the best work on the subject in any language. The last edition contains many beautifully executed and truthful engravings, and a very copious alphabetical index—so that you need only read what you wish to know at first; this index is important, as the book contains about 400 pages. After a while, however, you will delight in reading the book carefully through, from cover to cover. (E. P. 1861)*

Langstroth Hive

The Patent Wars

The 19th Century was a period of intense inventiveness in the United States, as well as the rest of the world. In particular, there was a spike in the number of patents which rose steeply after 1840 to peak around 1860. Beekeeping attracted inventors who generated dozens of new and varied "patent hives," each boasting of the supposed improvements, and all of them hoping to "get rich quick" riding the crest of the beekeeping boom of the period.

> *Farmers have probably been more generally disappointed in the purchase of patent hives than in any other article. In the purchase of patent machinery, or almost any thing, you can see it tested and judge of its merits before purchasing; but with hives, the purchase is mostly and of necessity made on faith, … he invests and waits patiently a season or two for the fulfillment of those promises, and then almost invariably regrets his outlay. (E.P. 1859)*

In fact, some doubted an "improved hive" was even necessary:

> *All experienced bee-keepers will agree, that as good results have been obtained, in many instances from the common box hive, or even from a section of a hollow log, with a surplus honey box or boxes, as from any patent in existence. All we can, beyond this, reasonably hope for, is to attain more uniform results. (E.P. 1859)*

Recall that beekeeping in the 1800s was generally carried out in the tradition of many centuries where bees were kept in the most primitive fashion: housed in baskets, boxes, or logs. These "patent hives" claimed a multitude of improvements, but in the end contributed little to the advancement of beekeeping and caused a great deal of discouragement and disgust among those who bought the "patent rights." It was in this climate that Langstroth brought forth his invention. He wrote:

> *In the present state of public opinion, it requires no little courage to venture upon the introduction of another patent hive, and an entirely new system of management; but I feel confident that a new era in bee-keeping has arrived, and invite the attention of all interested, to the reasons for this belief. A perusal of this Manual, will, I trust, convince them that there is a better way than any with which they have yet become acquainted. (Langstroth 1857)*

REV. L. L. LANGSTROTH'S

MOVABLE COMB HIVE.

PATENTED OCTOBER 5, 1852.

This hive gives the Apiarian perfect control of every comb, cell, and bee in the hive.

COMB REMOVED.—Each comb in this hive is attached to a separate movable frame, and may be all taken out in a few minutes, without cutting or injuring them in the least, or at all enraging the bees.

STOCKS STRENGTHENED.—By this arrangement, weak stocks may be easily strengthened, by helping them to combs, honey, or maturing brood taken from strong ones, and queenless colonies saved from certain ruin, by giving them the means of obtaining another queen.

FERTILE QUEEN—MOTH.—As all the stocks in the apiary, by the control of the combs, can be kept strong in numbers and in possession of a fertile queen, the ravages of the bee moth may be effectually prevented.

SEE WHAT IS WRONG.—If the bee-keeper suspects that anything is the matter with a hive, he can open it, and by actual examination of its combs, ascertain in a few minutes its true condition, and thus apply, intelligently, the remedies which it needs.

NEW COLONIES.—New colonies may be formed in less time than is usually required for hiving natural swarms; or the hives may be managed on the common swarming plan, or enlarged (without any alteration of existing parts), so as to afford ample accommodation for a non-swarming stock.

DRONES AND QUEEN.—By a very simple arrangement, the queen may be confined to her hive, while the workers have their liberty, so that bees may be left at any time, without the least risk of their swarming in the absence of the bee-keeper. The drones, when in full flight, may, by the same device, be excluded from the hive and destroyed.

SURPLUS HONEY.—The surplus honey may be stored in an upper box, in frames so secured as to admit of safe transportation—any one of which may be taken out separately and disposed of; or, if preferred, it may be stored in small boxes or glasses, in convenient, beautiful, and salable form.

TRANSFER COLONIES.—Colonies may be safely transferred from any other hive to this, at all seasons of the year, as their combs, with all their contents, can be removed with them, and easily instuned to the frames; and if this operation is skillfully performed in the gathering season, the colony, in a few hours, will work as vigorously in the new as they did in the old hive.

NO BEES KILLED.—If the combs of the bee-hive can be easily removed, and with safety both to the bees and the operator, then every enlightened bee-keeper will admit that a complete revolution must eventually be effected in the management of bees.

This is the original "Movable Comb Bee-Hive," and has been in use about nine years—insufficient length of time to test its value—and has been adopted by the most extensive practical bee-keepers. This hive is so eminently calculated to meet all the wants of the practical bee-keeper, that numerous infringements in the shape of some slight alteration are being vended by designing men, who well know that no patent hive can receive any considerable share of attention, that does not involve the movable comb principle.

TESTIMONIALS.

REFERENCES.

Persons desiring information in regard to the practical workings of this hive, are referred to the following gentlemen, as prominent among the thousands who have used it for years, and who are among the best practical Apiarians in this country:

STATE, COUNTY, TOWN, AND INDIVIDUAL RIGHTS.

An individual or farm right to use this invention, including one hive, will be sold for ten dollars. Such a right entitles the purchaser to use, and construct for his own use, on his own premises, any number of hives.

Applications for individual and territorial rights, in Maine, Vermont, New York, Ohio, Indiana, Michigan, Illinois, Wisconsin, Iowa, Missouri, Tennessee, California, and the Territories, addressed to the undersigned, will receive prompt attention.

R. C. OTIS, Kenosha, Wis.

Langstroth Flyer

In his *Reminiscences*, begun in 1893, Langstroth tells of the moment when his was so absorbed with his hive, with its moveable frames, the ease by which he could handle the bees, how he could rearrange the hive at will – that he failed to hear or even notice that there was another watching him intently. This other beekeeper was utterly astonished by what Langstroth was doing that he began to shout:

> *"Friend Lorenzo, you are so taken up with your new hive that you seem unable to hear me, or to see anything else. No doubt you think you have made a great invention: but I say you have made no invention,"* and then, repeating the words, *"you have made no invention. Friend Lorenzo, you have made no invention at all, but rather, a perfect revolution in bee-keeping!" (Langstroth 1893)*

But alas, others saw the merit of the new hive with its moveable frames, and began to design and sell "patent hives" of their own, which would embody the principle of the new hive – the bee space surrounding the frame, which prevented the bees from attaching the frame to the hive – but they would vary the dimensions or some other aspect to avoid infringing on Langstroth's own patent. As Annie Betts, the editor of the publication *Bee World* wrote in 1949: "It is the bee-space, and not the frame, that is the essential part of Langstroth's invention."

In the summer of 1852, Langstroth made more than a hundred movable-frame hives, some of which he sold. Most of the hives, however, were used in his own apiary. But a recurrence of his psychological disability incapacitated him and he was forced to sell his bees (Smith 1948). These periods of illness lasted months or even years at a time.

By 1861, the *American Bee Journal* had been formed by Langstroth's good friend Samuel Wagner, and was dedicated in part to the promotion of the beekeeping revolution. It also gives us a recording of the bitter rivalries between the various inventors and entrepreneurs who were determined to ride the tide and profit from the flood of inventions including the new hives.

Many individuals sold variations of these inventions and claimed full rights to the patent royalties. Further, a heated debate ensued as to whether the Langstroth hive itself was original or simply just a variation of "frame hives" devised during preceding decades. Thos. Wm. Cowan, editor of the *British Bee Journal*, described the situation like this:

> *There are not many bee-keepers of the present day [1895] who can look back 40 years or who know how Mr. Langstroth was treated, even by those who were quick to perceive the advantages to be derived from his invention ; or how they pilfered his best ideas, and even patented them, and how he was defrauded of his just dues. Nor do they know that these infringements of his rights led to costly litigation which swallowed up all his well-merited gains. As Prof. Cook has written in Gleanings: "This whole matter is the dark page in American bee-keeping history, and we gladly pass it by without further comment."*

The Italian Bee

In his moments of lucidity, Langstroth clearly saw the future direction that beekeeping would take:

> *April 9th 1852. I am disposed to place my plan for rearing young queens as next in importance to movable frames – the removal of the old queen for one week will give time for sealed young ones – her removal for so short a time will be of no serious detriment to the strong old hive – When hives swarm let all the sealed young queens be at once removed and a young hatched one at once given – this will save a week or more of precious time. (Langstroth)*

The dream of a "better bee" had already been kindled; at the same exact time, the pioneering geneticist and monk Gregor Mendel was importing bees from around Europe to his apiary in Moravia, in today's Czech Republic. According to Vecerek (1965). Greqor Mendel imported yellow bees from Cyprus and intended to use them in his experiments to discover the genetic basis of traits in plants and animals. Similarly, the golden Italian bee would present a sharp contrast to the prevalent European black bee, then in general use in the United States. Langstroth was immediately taken by the idea of bringing Italian bees to the United States and giving them wide dissemination, so he became closely involved with the effort.

> *I can probably give, better than any one living, the history of the first efforts made to introduce Italian bees into this country; as I knew well the late Messrs. Samuel Wagner and Richard Colvin, and Messrs. S. B. Parsons and P. G. Mahan, who, with myself, were the first to import them. (Langstroth 1881)*

More recently (1969), Toge and Mildred Johansson published a long article in *The Journal of Long Island History* with the tantalizing title "Samuel Bowne Parsons and the Golden Bees." Parsons was a nursery man in Flushing, Long Island, which served as the heart of many horticultural enterprises set up to supply the voracious appetite of Americans for exotic plants from all over the widely explored world. His accomplishments were numerous and included the importation of the beautiful Weeping Birch and the penning of *The Rose; its History, Poetry; Culture and Classification*, in 1847. His son Samuel B. Parsons Jr. continued the family tradition by becoming a landscape architect and the Superintendent of Planting in New York City (1881– 1910) and oversaw the plantings in Central Park. Following the Johansson's story, we read:

> *Beekeepers had long been tantalized by descriptions in Virgil's Georgics of a bee "glowing with spots of gold," the Italian race of the honey bee Apis mellifera Linnaeus. Captain Hauptmann von Baldenstein observed these bees in Northern Italy during the Napoleonic wars and later (September, 1843) arranged to have a colony carried across the Alps to his home in Switzerland. (Johansson 1969)*

By 1860, after many setbacks, Parsons had a few live colonies of Italian bees in Flushing, NY Since he knew nothing about beekeeping, he enlisted the help of Rev. Langstroth in saving what was left of his ill-fated attempts to import bees. Of twenty that were shipped, but one was left alive and in a weakened state. A subsequent shipment succeeded in arriving with two more live queens. Parsons wrote: "I have therefore now only three queens from the sum of eight hundred dollars." He evidently anticipated making a lot of money on the enterprise, perhaps repeating his success in riding the "silk culture craze" by planting 25,000 mulberry trees and launching his career as a nurseryman in 1838 (Johansson 1969). Langstroth and his contemporaries succeeded in developing the practice of propagating Italian queen bees and shipping through the mail, a practice which is still widespread at the present and comprises millions of queen bees annually.

Final Words

The Rev. L. L. Langstroth continued to innovate, by 1868 devising and selling centrifugal honey extractors, with the aid of his son who assisted him in business. During the periods when his mental illness rendered him unfit to work, his wife helped him with revising his book, and revising and expanding his patent. Sadly, he outlived both his wife and son, and died at the age of 84 while giving a sermon. After his death, there was an outpouring of appreciation of the man, his accomplishments, and his wonderful personality. The journal *Scientific American* confidently proclaimed that:

> *The Langstroth hive and frame created the first and probably the greatest revolution in bee-keeping. The movable frame is a device by which the inside of the hive can be removed by sections, and without disturbing the bees in other parts of the hive … The modern method of introducing Italian queen bees in the hives is probably fully as important in its results as some of the foregoing inventions, and it should be classed high among the factors that have brought larger profits to the apiarists in all parts of the country. (Walsh 1899)*

LORENZO L. LANGSTROTH
1810 – 1895
PASTOR OF THE SOUTH CHURCH
ANDOVER, MASSACHUSETTS
MAY 11, 1836 – MARCH 30, 1839.
ERECTED IN THE CENTENNIAL YEAR OF HIS DISCOVERY OF THE BEE SPACE AND HIS INVENTION IN 1851 OF THE MOVEABLE FRAME WHICH MADE MODERN BEEKEEPING POSSIBLE.
DEDICATED JULY 22, 1951
BY
THE MASSACHUSETTS FEDERATION OF BEEKEEPERS' ASSOCIATIONS
IN RECOGNITION OF HIS OUTSTANDING WORK IN BEEKEEPING.

Langstroth plaque

Works Cited

Cowan, T. W. (1895). Langstroth Memorial. Gleanings in Bee Culture. 23:24. A.I. Root Co. Medina, OH.

E. P. (1859). The Country Gentleman. 14:9. Luther Tucker & Son. Albany, NY.

E. P. (1861). The Country Gentleman. 17:1. Luther Tucker & Son. Albany, NY.

Johansson, M. P. & T. S. K. (1969). Samuel Bowne Parsons and the Golden Bees. The Journal of Long Island History. Winter - Spring. Judd, Jacob.

Langstroth, L. L. (1852). Bee Hive, Patent No. 9,300. United States Patent Office.

Langstroth, L. L. (1853). Langstroth on the Hive and the Honey-Bee, A Bee Keeper's Manual. Northampton: Hopkins, Bridgman

Langstroth, L. L. (1857). A Practical Treatise on the Hive and Honey-bee. CM Saxton & Company.

Langstroth, L. L. (1881). Early Importations of Italian Bees. *American Bee Journal*. 17:11. p. 82. Langstroth, L. L. (1893).

Langstroth's Reminiscences: How he became interested in bees. Gleanings in Bee Culture. 21:3. p. 80.

Hubbell, S. (1998). A book of bees: And how to keep them. Houghton Mifflin

Harcourt. Naile, F. (1942). The Life of Langstroth. Cornell University Press. Ithaca, N.Y.

Phillips, E. F. (1951). Address at unveiling of plaque in South Church, Andover, Massachusetts July 22, 1951.

Smith, O. D. (1948). Langstroth: The "Bee Man" of Oxford. Ohio State Archaeological and Historical Society

Walsh, G. E. (1899). Life in a Winter Bee Cellar. Scientific American. 80:23 p. 380.

Vecerek, O. (1965). "Johann Gregor Mendel as a beekeeper." Bee World 46:3 p. 86-96.

Chapter 6

The Gold Rush and the Honey Bee

The Harbison Brothers

John and William Harbison were born in Western Pennsylvania in the early 1800s. Their father kept bees in rustic hives, so they were exposed to beekeeping at an early age. Before discussing their success in transporting honey bees to California, I want to describe the unique hive they used and popularized.

John Harbison   William Harbison

The typical hive of the period was little more than a wooden box. Some more ambitious beekeepers had learned to provide two compartments, one for the queen and the rearing of brood (developing bees), and another above this, for the storage of honey. These compartments were accessed by means of doors, one for each section. Often the beekeeper would arrange glass jars over holes in the shelf, and the bees stored their honey in the jars - not knowing that it would be subsequently removed by their keeper.

During this same period, Prokopovych (Прокопович), working in the Ukraine, developed and built hundreds of cabinet hives and it appears that he was the first to place wooden frames in the upper chamber of his hives, to obtain comb honey in small wooden boxes. This technique would be adopted by the Harbisons, but they took the hive one step further, by providing moveable frames in the lower chamber so that the combs with the queen and her brood could be examined and even removed in order to propagate bees in new hives, which the Harbisons called "colonizing." W. C. Harbison wrote:

> We claim an improvement in the mode of constructing, and using
> the frame— that is essentially different from any previous known

device, yet for a similar purpose for which frames have long been used. For which improvements, letters patent, were granted January 4th, 1859. (Harbison 1860)

He is referring to the fact that their hive is a cabinet hive where the frames can be removed from the back, individually, without disturbing the other frames. Previous hive designs had frames which were hung crosswise, so that the beekeeper had to remove all the other frames to access the one furthest inside. That the Harbisons were able to successfully move colonies of honey bees from the East Coast to the West Coast, via Panama, is in a large part due to the design of their hive. Many thousands of bee hives were shipped west beginning with the California Gold Rush boom, but most perished during the long voyage.

Harbison hives

It is interesting to note that while The Harbisons' hive gradually fell into disfavor, one very much like it is still considered standard in Slovenia, and neighboring countries. The hive, called the "A-Ž hive" in honor of inventors Alberti and Žnideršič, has back opening cabinet doors, and frames which are removed like books in a bookcase. The Harbisons were well aware of the newly patented Langstroth hive, but preferred their cabinet hive. Ironically, W. C. told a newspaper reporter from the *Dollar Newspaper* for January 21, 1857:

I will venture the prediction that both Quinby's hive and mine will ere long be cast aside, to give place to a hive constructed in such a manner that the apiarian can have access to every part of the hive at pleasure, without injury to the colony. In this particular both Mr. Quinby and myself have signally failed. The invention of such a hive was reserved for Mr. Langstroth. (Cook 1880)

The Harbisons Bring Bees to California

Growing up in southern California, I was exposed to Gold Rush lore at an early age. We diligently learned of the Forty-Niners, and their exploits. California gold was discovered in January of 1848, but it took a year before the news reached the East Coast, when President Polk made an official announcement in his State of the Union address to the U. S. Congress. (LOC 2020). The following year, some 100,000 people from around the world rushed to California. While a great many were engaged in the actual prospecting and mining of gold, there was also the huge contingent of people whose aim was to sell goods and services. California was viewed as the place to go to get rich quick. It turned out to have valuable resources other than gold, which figured in the success of people like the Harbisons. The central valley of the state was destined to become one of the most productive agricultural regions in the world, and the forests of virgin redwood trees constituted a find whose value exceeded the discovery of gold. For many decades, nearly everything wooden – from houses to hives – was made from redwood lumber.

Harbison Residence c.1885

One of the principal biographers of John S. Harbison, was Lee H. Watkins. He was apiary technician for the Entomology Department at U. C. Davis for many years, and wrote a lot about early California beekeeping. He was published in various magazines, including the *San Diego Historical Society Quarterly*. Sadly, Lee died unexpectedly in 1972, about 64 years old, leaving behind a great deal of unfinished work. His bibliography of works on Apiculture and Sericulture was completed by William Pickens in 1975. Watkins wrote this about John Harbison's California adventure:

> *It took only a few weeks of unrewarding digging near Campo Seco, Calaveras County, to convince him that gold mining was not for him. He then traveled to Sacramento and went to work on December 5 at the T. F. Gould and Company's sawmill in Sutterville, where he was employed until the fall of 1855. He also soon wrote East (probably his brother, William) for some seeds and a few trees, and when this arrived in February 1855 he opened a nursery on the Jefferson property in Sutterville. (Watkins 1969)*

Using their varied expertise (long distance traveling, beekeeping, lumber milling, nursery growing and general entrepreneurship), the Harbisons were able to do what many had tried and subsequently would fail to do: bring live bees to California where none were, and begin the rapid build-up of beekeeping and honey production on the west coast. An article in the *Sacramento Daily Union*, December 5, 1857, declared:

> *On Thursday afternoon last, the firm of W. C. & J. S. Harbison, shipped on board the Indiana, Capt. Laughlin, sixty-seven colonies of bees for California. J. S. Harbison, of Sacramento city, had them in charge; who, it appears has been here the past season arranging his hives in a proper manner, and having the bees lay in their stock of provisions for their long journey. Every comb in the hive is firmly supported in its place, and each hive is ventilated and covered with painted canvass, in order to protect the bees from heat and rain. These gentlemen have much experience in the bee business – in fact the pioneers in doing a heavy business in Lawrence county [Pennsylvania] and, if successful, will be the pioneers in the trade in California. This we believe is the heaviest shipment of bees ever made to the El Dorado; however, if we recollect, an attempt was made some time since, but failed, in consequence of not having them properly packed to pass through a tropical climate. A colony of*

> bees in California, we are told, is worth about $100, and thus we may
> judge that if these gentlemen succeed with this shipment they may
> soon realize a fortune with their bees. (Anon 1857)

Sherman Was There

In the course of my research into the Harbisons, I stumbled upon a fascinating series of articles, published in the *California Historical Society Quarterly*, beginning in 1944 and with the title "Sherman Was There: The Recollections of Major Edwin A. Sherman." Here was another ambitious and enterprising man, destined to cross paths with the Harbisons. While the Harbisons' books ring true, and are filled with information of the time, Sherman wrote considerably more than they did, and by his own account, was the actual writer of John S. Harbison's *Bee-keeper's Directory*. In his words:

> August of 1859, I received from a friend a letter of introduction to a
> Mr. John S. Harbison, who lived near the Sacramento River about four
> miles below Sacramento City, and who, I was informed, wanted some
> capable person to write his book on "Bee Culture." (Sherman 1945)

According to Sherman, he apprenticed with Harbison for four months, working during the day and writing at night. He states that hives were still selling for $100, but the shrewd Harbison knew that if the market began to saturate, prices would plummet, so Harbison persuaded Sherman to become an agent and sell hives in a part of California where they were still lacking, namely Los Angeles. He soon learned that spring arrives two months earlier there than in Sacramento, and by March the hives were filled with bees and preparing to swarm. Central to Harbison's system was to multiply the colonies at will, *before* they issued swarms, which would simply head for the hills.

As luck would have it, Sherman was called to jury duty at the point at which the most work needed to be done – forming new saleable colonies – before the swarming fever took hold and caused his investment to be lost. He wrote that the local Sheriff was so adamant about Sherman's duty as a juror, that he didn't even have time to wash his hands. These were literally coated with the sticky aromatic resin that bees paint all over the inside of beehives. Following his story, the smell of the resin was so attractive to the bees that they pursued him to the court house and proceeded to sting the Judge, the jury and worst, the horses tied up outside. The trial was dismissed and our beekeeper led his bees out of the courthouse, now a hero.

Now, I don't know how much credence to place in this tale but Sherman states that he took the proceeds from the sale of hives to purchase a printing press and a newspaper: The San Bernardino *Herald*. This he renamed the *Patriot*, and he referred to it as a "a staunch uncompromising Union journal, loyal to the government and the American flag." His story continues with "The Discovery of Silver in Nevada," and from there we say farewell and bid him luck.

Beekeeping Takes Off in California

John Harbison's California beekeeping career began in the vicinity of Sacramento, which is located at the southern part of the Sacramento Valley. It merges imperceptibly with the San Joaquin Valley directly to the south, forming what was referred to as the Great Valley, or more commonly, the Central Valley. It comprises over a third of the acreage of California. The value of this region was immediately recognized and in a few years the vast meadow was turned into farmland, leaving about one percent of the native grassland untouched. Unfortunately, this happened so quickly that there is a little record of the original botany of the region, and it was quickly replaced by farm crops and their associated weeds.

> The settlement of California by white people has been so recent, and the peculiarities of its flora and fauna so different from most other parts of the world, that the introduction of many plants and animals common elsewhere, but not indigenous here, has occurred within the memory of men of this generation, or the one preceding. (Fox 1878)

I have been unable to find out what made up the original flora of the Great Valley, but it was primarily grasses - hence the fact that it was quickly plowed up and sowed to wheat. Some beekeepers apparently thought honey came from the tall "tule grass" (*Schoenoplectus*) which grows up to ten feet high in some places. One writer stated:

> The tule has been mentioned as a honey plant since many beekeepers claim it to be one, but the writer believes that it yields no nectar. Honey buyers often refer to honey gathered along the Sacramento and San Joaquin rivers as "tule honey." In marshes about Sacramento River. (Richter 1911)

The Central Valley marshes had many species of plants, but the main source of honey was probably *Cephalanthus*, also known as buttonbush.

> In the Sacramento and some other valleys in California, the *Cephalanthus* abounds along streams of water or in the edges of the Tule lands, where it grows very large and yields immense quantities of honey, of the best quality in the State, and scarcely inferior to any in the world. (Harbison 1860)

There are also many references to wild mustard (*Brassica spp*) as a honey plant:

> The California honey, made from mustard blossom, the flower from which most of the honey is gathered in this valley, is equal to any I have ever tasted. It has sold in San Francisco at from $1.25 to $1.50 per pound. (Appleton 1858)

Harbison

Even Harbison refers to it in his 1861 *Bee-keeper's Directory*, stating that "Mustard affords a larger amount of valuable pasturage to the acre than almost any other plant. It blooms throughout the month of May, and part of June."

It is difficult to know when the flora of the Valley was no longer predominantly native, and when it had become mostly crops and weeds. Many of these had been imported earlier by Russian settlers who established Fort Ross about 1814, a hundred miles or so west of the Sacramento Valley. They grew a variety crops including wheat and barley as well as vegetable such as potatoes, beets, cabbages, and also mustard.

The Move to San Diego

John Harbison was not to remain in the Sacramento area. He was a sharp businessman and had already realized that if he was successful in making his fortune by supplying the local settlers with bees and hives, eventually everyone would have them and the market would collapse. As mentioned, he had agents plying the trade in other counties. In this way, he was not only looking for new opportunities to sell hives and bees, but he sought to find new untapped pastures where he could produce honey himself, which still commanded a high price.

The Great Valley periodically flooded and finally, "all his earthly belongings were lost in the Sacramento Valley in the flood of 1862." (Hanson 1923) Harbison had learned of the bee pastures of San Diego County, where the hills and mountains abounded with species of sage (*Salvia*) and buckwheat (*Eriogonum, Polygonum*). He rebuilt his apiaries from those that had escaped the flooding, and in 1869, moved his holdings to San Diego. His knack for selling Harbison Hives seemed unbounded and as a result by 1876, census records showed over 23,000 colonies of bees in that county. (Fox 1878)

The result of this explosion in beekeeping was inevitably a glut of honey on the local market. Undaunted, Harbison took advantage of the newly completed (1869) transcontinental railway which was shipping products back and forth between California and the Eastern United States. The *American Bee Journal* of November, 1873 reported:

> *Clark & Harbison, of San Diego, Cal., have made quite an extensive shipment of honey to this city [Chicago, IL]. They sent one car-load containing 21,000 pounds comb honey, which is the largest shipment ever made at any one time to this market. Mr. Harbison informed us that they had obtained over 60,000 pounds comb honey this season from their southern apiaries alone. (Anon 1873)*

In the following years he produced over one hundred such freight car loads of honey. Harbison's California Sage Honey could be purchased in stores all along the east coast. At the time, the standard package for the honey consumer was a wood box with a glass cover, holding five pounds of honey. Harbison introduced smaller one-and two-pound "sections" which were more attractive to the consumer, and quickly became the standard package for comb honey.

Eventually, the honey extractor was invented and widely adopted. The demand for liquid honey surpassed that for honey in the comb. Harbison never made the switch from comb to liquid honey. Instead he reduced his holdings in bee hives and followed other lucrative pursuits.

> By the late 1880's Harbison had considerable investments in real estate and orchards himself, as well as being active in the Harbison Wholesale Grocery Company. Though he had 500 colonies of bees in 1893, most of them were rented out. One hundred colonies were still in his possession when he died [in 1912]. (Watkins 1969)

Beekeeping in San Diego County struggled against many set backs. One of these was the notion of fruit farmers that honey bees were detrimental to their crops. Not only did they not realize the need for bee pollination but they were convinced that honey bees ate the fruit. Even when court testimony revealed wasps as the real culprits, fruit growers burned whole apiaries in retaliation.

Harbison Canyon

Harbison and the Italian Bees

The Harbisons expanded their influence throughout California by importing bees from the East and manufacturing hives by the thousands. Also, both John and his brother William published full length books on bee biology and the method of employing their hives. Like those of Langstroth, Quinby and others, these books were widely read and absorbed by beekeepers.

Following the importation of the Italian bee to New York by Langstroth and Parsons, this breed was rapidly adopted by progressive beekeepers, in part because of its productivity, ease of management, etc., but also because of its attractive golden appearance. Not to miss out on another opportunity, John Harbison obtained Italian bees to propagate and sell to the willing beekeepers in California, that they might convert to superior stock.

> In 1865 he went east, and on his return brought with him selections of Italian bees, made from the most noted apiaries in the eastern states, and also seventeen of the choicest queens he could find. The hives of Harbison are now composed entirely of the very best honey making bees. (Anon 1873)

In a very real sense, Harbison was the primary driver of apiculture in California, although it is plain had he not done it, others would have. But his efforts were so strong and successful, that he earned the title "King of the Beekeepers." In the hills east of San Diego a deep valley bears the name "Harbison Canyon." Like much of California, it is a high risk area for wildfires. The slopes are covered with the black button sage and wild buckwheat that filled Harbison's hives and those of countless beekeepers in the decades since. When the summer sun scorches the chaparral, the scrubby bushes are fire waiting to happen. But when the winter rains come and soak their roots, the hills come alive with blossoms and the honeycombs fill with nectar.

Works Cited

Anon. (1857). Honey Bees in Sacramento. *Sacramento Daily Union*. Vol. 14, No. 2089.

Anon. (1873). Apiaries in San Diego, Cal. *American Bee Journal*. Vol. 7, No. 9.

Anon. (1930). Old Fort Ross - A Stronghold of the North. *California History Nugget*. Vol. 3, No. 3.

Appleton, F. G. (1858). *California Farmer and Journal of Useful Sciences*. Vol. 9, No. 8.

Cook, A. J. (1880). *The Bee-keepers' Guide: Or Manual of the Apiary*. Chicago, Ill., T. G. Newman.

Fox, C. J. (1878). Bee-keeping in California. *Bee-keeper's Magazine*. Vol. 6, No. 11.

Hanson, F. (1923). How Honey Bees First Came To California. The Western Honey Bee. Vol. 11. No. 1.

Harbison, J. S. (1861). *Bee-keeper's directory; or, The theory and practice of bee culture*. H.H. Bancroft.

Harbison, W. C. (1860). *Bees and Bee-keeping: a Plain, Practical Work*. CM Saxton, Barker & Company.

Library of Congress. (2020). The Discovery of Gold. www.loc.gov/collections/california-first- person-narratives/articles-and-essays/early-california-history/discovery-of-gold/

Richter, M. C. (1911). Honey Plants of California, Bul. 217. University of California, Berkeley.

Sherman, A. B., & Sherman, E. A. (1945). Sherman Was There: The Recollections of Major Edwin

A. Sherman. *California Historical Society Quarterly*. Vol. 24, No. 3.

Watkins, L. H. (1969). John S. Harbison: California's first modern beekeeper. *Agricultural History*. Vol. 43, No. 2.

Watkins, L. H. (1969). John S. Harbison: Pioneer San Diego Beekeeper, *The Journal Of San Diego History, San Diego Historical Society Quarterly*. Vol. 15, No. 4.

Chapter 7

Samuel Wagner and The *American Bee Journal*

Bees, sheep, and angle-rod, be sure,

Will make thee quickly rich—or poor!

In his 1938 book, *The History of American Beekeeping*, Frank Pellett wrote: "The history of the *American Bee Journal* has been the history of the rise of beekeeping, and the one is inseparably linked to that of the other." This is so, but also true is that history is incomplete without Samuel Wagner.

Samuel Wagner was born in York, Pennsylvania, in 1798. His parents were German immigrants. Samuel spoke only German until he was ten, and was fluent in both languages as an adult. At 26, he purchased a newspaper, *The York Recorder*, and became its editor. Evidently, he kept bees on the side, and had been reading German beekeeping publications. Sometime around 1850, he became acquainted with L. L. Langstroth, the inventor of the hive which bears his name and which became adopted throughout the world as the best for successfully managing bees.

Wagner was studying the work of Jan Dzierżon, who was advancing the state of beekeeping in Europe. Dzierżon stated that though he had been born in Poland, "I became a German by education. But science does not recognize borders or nationality." Wagner had completed a translation of Dzierżon's book *Rational Beekeeping*, and was hoping to publish it in the US. But when he learned of Langstroth's work, he felt that this would make a better book. He offered his services to help Langstroth produce what became *The Hive and the Honey Bee*, in 1853. Wagner turned his attention to a monthly journal for beekeepers, inspired by Dzierżon's own *Der Bienenfreund aus Schlesien* (The Silesian Bee Friend), which began publishing in 1854.

The early history of the *American Bee Journal* in was described in detail by editor George York in 1903. In his words:

> *I have told you of the value to be derived from the early works of some of the writers on this subject, [but] it is impossible to keep abreast with current developments of the bee industry and confine*

yourself to text-books alone. To supply this deficiency the need for a bee journal was felt; and in January, 1861, Samuel Wagner started the publication of the American Bee Journal at Philadelphia, Pa. (York 1903)

I think it's worth including here a brief quote from Gardner's *The Rise and Fall of Early American Magazine Culture*:

Magazines were far from ephemeral. Unlike newspapers—which, then as now, were rarely preserved once the day's news had been consumed—the individual periodicals were often carefully saved, collected so that they might be bound at year's end into a handsome volume. These were items to preserve for posterity as well, and we can trace in the signatures that grace the title pages of bound volumes the generations through which many of these magazines passed. (Gardner 2012)

The First Volume

The first issue of the *American Bee Journal* was filled with material chiefly authored by Samuel Wagner. Writing in 1861, he advised the reader that failure with beekeeping is attributable to three main factors: lack of basic understanding of honey bee biology; hives that are poorly made and/or poorly designed; and the absence of skilled and knowledgeable management. He added that in the previous decade crucial discoveries had been made and information needed was already available, albeit as articles in various agricultural periodicals as well as pamphlets and books, but perhaps was not widely distributed. To turn the situation around, he declared his intention to produce a journal dedicated to all aspects of beekeeping and honey production. He wasn't about to merge it with poultry production, like Thomas Miner did for his magazines.

Wagner told the reader:

It is not proposed to give the Bee Journal a predominantly scientific cast. Aware that to be extensively useful, it must adapt itself to the wants of the community, it will constantly regard that object. Its contents must be diversified. Its columns must be accessible alike to the apiarian, whose experience and observations enable him to communicate information, and to the inquirer whose primary desire is to obtain instruction. (Wagner 1861)

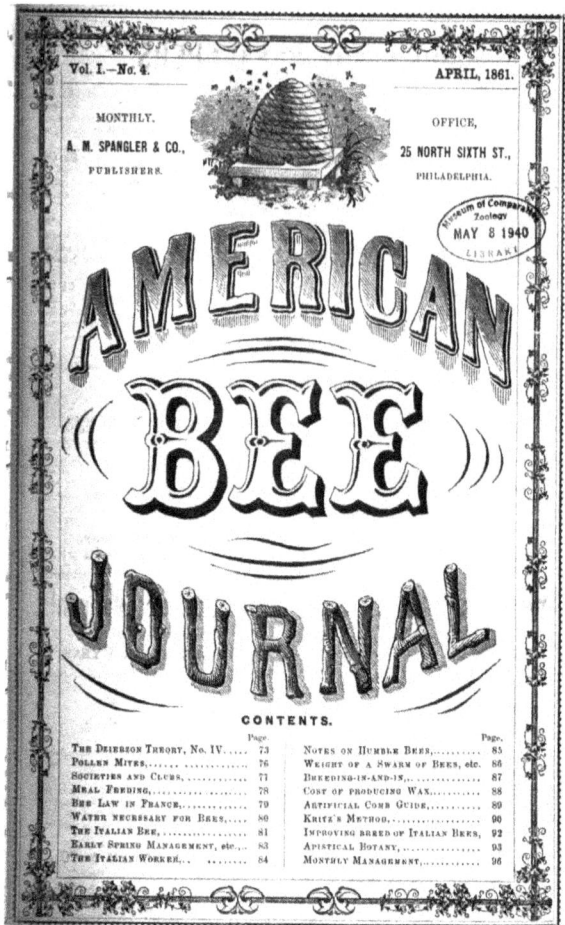

American Bee Jounal Cover 1861

Wagner took it upon himself in this Issue Number One, to fully explain what he called "The Dzierżon Theory." He outlined thirteen basic principles that were considered to be essential to the theory, and which had been confirmed through careful experiment and observation. Much of this was far in advance of the understanding of genetics and biology that we take for granted today. Central to Dzierżon's teaching was that an unfertilized honey bee egg will develop into a male (drone) bee; that the queen is inseminated outside the hive at the beginning of her life; and that she fertilizes within her body all the eggs which are destined to become females.

Further, that while the majority of these females will be ordinary worker bees, the workers themselves can transform a female larva into a queen by special feeding. These ideas were not universally accepted at first; Wagner added: "With the annunciation of Dzierżon's theory, a new era commenced; and though that theory encountered warm opposition, and excited a protracted controversy, it has triumphantly sustained itself, and led to further important discoveries." One of his readers called out his lengthy explanation of Dzierżon's discoveries but referred to them as "The Wagner Theory." His response:

> No full and precise statement of what is known as the "Wagner theory" has ever been made. The brief reference is a mere outline of its general features as then held by us, subject to such modification as further observation and reflection might suggest. We have not since felt called on to present our views more in detail, nor do we propose to do so now. (Wagner 1861)

The first issues of the journal were not confined to descriptions of biology, as Wagner was keen to impart information of a practical nature. He clearly explains when and why bees sting, and when the beekeeper should avoid meddling with their affairs: "In the cool of the morning, in the dusk of the evening, and during rainy weather." Wagner advises that the bees be kept under the beekeepers control by the proper application of smoke while at the same time avoiding those actions that especially irritate them, including:

> ...improper treatment or rough usage. Rapid motion in front of the hive, obstructing the flight of departing or returning bees, breathing on them when clustered, and beating against their hive or its bottom board, are annoyances which excite their ire, and should be carefully avoided. Various odors, the natural exhalations of some persons, and the effuvium of their own poison when discharged, rouse their animosity and inflame their rage. (Wagner 1861)

Even so, Wagner clearly declares that the "genuine apiarian must have no fear of the bee's sting" to which his system will soon become accustomed and rarely will it produce serious consequences. Fearlessness is key to success with bees. (Advisory: experienced beekeepers will also tell you that serious accidents can take place, leading to episodes of severe stinging.)

Initially, the ABJ ran about 30 pages per issue, including four or five pages of advertising. The December 1861 issue, however, was only 20 pages with only one page of beekeeping ads. The back cover featured a full page declaring the virtues of "The Garden State of the West." They offered for sale "1,200,000 acres of rich farming lands;" not in California but Illinois!

Unfortunately, the situation in the country had profoundly changed. Wagner announced:

> TO OUR FRIENDS. With this number (which has been somewhat delayed from unavoidable causes), we conclude the first volume of the American Bee Journal and now announce that the publication will be suspended for a year, to be then resumed if the state of the country will admit, and those interested in bee culture desire it.
>
> The first volume of the American Bee Journal is now completed, and will be sent free of postage on receipt of One Dollar. (Wagner 1861)

The Civil War Intrudes

"The war between the States" had begun and it would not be until 1866 when the journal would resume publication. By then Wagner had moved to Washington DC, where he gained employment as "disbursing officer" of the U. S. Senate. He retired in 1868 to devote his energy to his journal.

In May of 1865, President Johnson declared the Civil War to be "virtually at an end," although it would be in August, 1866 when he signed a "Proclamation – Declaring Peace." Due to the changed circumstances in the country, readers were told:

> The greatly increased cost of paper and printing constrains us to advance the price of the American Bee Journal to two dollars a year. Nevertheless we feel assured that it will be found "cheap at that," by every practical bee- keeper. The amount of original matter which each successive number will supply—most of which could not be obtained from any other source, or through any other channel, and all of which is thoroughly scrutinized before insertion, will be a rich equivalent for the enhanced price. (Wagner 1866)

The *American Bee Journal* recommenced with the July 1866 issue: Volume II, Number 1. It began with a lengthy essay on the circumstances which brought the honey bee to the Americas, the various debates as to when and how, and so forth. This was immediately followed by a discussion of a practical nature: in what direction should hive entrances be placed.

American Bee Jounal Cover 1866

Wagner lays down plainly that although beekeepers' prejudice may dictate that the openings be pointed south or southeast, "actual experience warrants me in saying that the point of the compass towards which the entrance to the hive is turned, is not of the least possible importance." He does, however, instruct that: "No long grass or tangled weeds of any kind, no cabbage or lettuce plants should be suffered to grow within two yards of an apiary, more particularly in front," and "The vicinity of lime or brick kilns, tan-yards, gas-houses, and offensive premises of every kind, is annoying to bees."

Other Projects

The same year he established the magazine, he obtained a patent for the manufacture of "honeycomb foundation." This idea was nearly as revolutionary as Langstroth's idea of putting bee combs into a moveable frame which could be standardized and made interchangeable with any hive in an apiary, and ultimately - any hive in the country, as it is today.

Given that it is sometimes difficult to induce honey bees to build a straight and regular comb in an empty frame, many experimenters tried to come up with an answer to this problem. Leaving aside the various false starts, the ideal solution was to prefabricate a sheet of beeswax with the basic foundation of the honeycomb stamped onto it, either by engraved plates or rollers, using ordinary printing methods. Impression of the cell bases could even be created with metal slugs like those used for individual letters in a printing press.

Honey-Emptying Machine.

Honey-Emptying Machine

L. L. LANGSTROTH & S. WAGNER.
APPARATUS FOR EXTRACTING HONEY FROM THE COMB.
No. 61,316.   Patented Jan. 15, 1867.

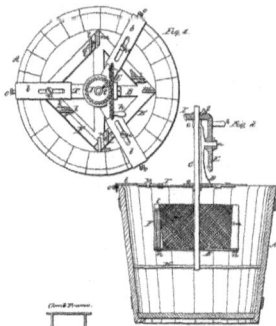

Langstroth Wagner Patent

Wagner was not the originator of this idea. The credit for that is usually given to Johannes Mehring of Germany, in 1857. In writing his patent, Wagner improved upon Mehring's idea which was to stamp flat diamonds into the wax. Wagner proposed that by leaving gaps between the cell base stamps, slight ridges would be formed giving an exact replication of the bees' manner of building out the cell from the comb's midrib. Oddly, he had no means to mass produce the product. According to Frank Pellett: "his patent probably delayed perfection of the process in the United States by others." Several decades were to pass before "honeycomb foundation" would be perfected and marketed, principally through the efforts of Amos Root and Edward Weed, who patented their own processes in 1896-1900.

Additionally, Wagner and Langstroth patented a honey extractor in 1867, based upon the ones appearing in Europe at the time. Their version is a heavy wooden tub, equipped with a crank and gears to spin honey out of the combs. Clearly their idea was not new, though the use of gears appears to originate with them. As a matter of fact, Langstroth states in the *American Bee Journal*: "Any one is, of course, free to make them." Other inventors soon followed with their own versions of honey spinners. As early as 1869, H. O. Peabody patented his, with a tank made of much lighter sheet metal, such as honey extractors are made today.

The state of beekeeping in Europe was far in advance of the U.S. in the 1850s, which Wagner knew from reading the German periodicals. So, he quickly saw the imperative of importing the Italian bee. In 1855, Wagner and Edward Jessup, of York, Pennsylvania, first attempted to get Italian bees to America. The bees evidently starved *en route*. Again, success fell upon others: S. B. Parsons, a nurseryman doing business in Flushing, NY, was selling Italian queen bees by 1861. In fact, the first issues of the *American Bee Journal* had numerous advertisements offering Italian bees for sale. They could also be bought from Richard Colvin, Baltimore; C. Wm. Rose, of Hempstead, Long Island; even Langstroth, now in Oxford, OH.

Wagner himself appears not to have gotten deeply into the queen business, as evidenced by his statement in the *American Bee Journal*: "We avail ourselves of the opportunity to add, in reply to frequent inquiries, that we have not been, and are not now, breeding Italian queens for sale, and have no pecuniary interest in any that are so bred."

The *American Bee Journal* became central to his life. In its earliest pages one could find extremely detailed expositions on anatomy, behavior, and physiology. These were interspersed with ongoing debates. From an 1868 issue:

> The Bee Journal is an arena where all ideas and opinions can meet
> and struggle. The common sense of the public will, sooner or later,
> judge without appeal, and decide in favor of the true and the right.
> A thing valued as good to day, may be rejected to-morrow, and
> replaced by something better. All this is very plain, and fair.
> (Wagner 1868)

Debated endlessly were the relative merits of different hives. In one such article, the author names Davis' Platform hive, Langstroth's Movable Frame hive, the American hive (Harbison's), the Quinby hive, Flander's Triangular and Hoop-frames hive, and also his Book hive. He declares them all lacking and objectionable, and proposes a new hive, which he had conveniently patented (Conklin's). Mr. Wagner stood above the fray and let the public decide for itself the relative merits of the hives, principles of biology and of management.

Samuel Wagner's Death

Rev. Langstroth, Wagner's longtime friend and associate, was saddened to have to write the following words in the March, 1872 issue.

> *READERS OF THE BEE JOURNAL :– Your dear old friend, the honored editor of the American Bee Journal, is dead. On Saturday, February 17th, he awoke early, partially dressed himself, and was talking pleasantly with his wife, when he was suddenly seized with shortness of breath, soon became unconscious, and in less than fifteen minutes breathed his last. The physician pronounced his disease to be aneurism of the heart. (Langstroth 1872)*

Wagner Portrait

Langstroth declared him "Better acquainted with the history and literature of bee-culture than any man in America, perhaps than any living man—seldom if ever forgetting a single fact once lodged in his extraordinary memory; he was so modest and reserved, that only those who knew him well, understood the wide range of his reading and investigation." A subscriber wrote the following letter, upon learning of his death:

> *He was respectful and courteous to those he deemed honest, and his criticisms, though often severe, were just, and although modest, he shrank not to expose, with cutting words, the noisy drones and pretenders in our great human hive. He certainly was a good*

judge of human nature. To benefit his fellow men seemed to be the bent of his mind. He did not live for the present alone, and many generations will have come and passed away before the name of Samuel Wagner will be forgotten. – W. P. H., Murfreesboro, Tenn., March 14, 1872. (Wagner 1872)

Albert Cook, wrote in his *Bee-keepers' Guide: Or, Manual of the Apiary* (1884) describing the various periodicals of the time:

With what pleasure I remember the elegant, really classic, diction of the editorials, the dignified bearing and freedom from asperities which marked the old American Bee Journal as it made its monthly visits fresh from the editorial supervision of Mr. Samuel Wagner. Someone has said that there is something in the very atmosphere of a scholarly gentleman that impresses all who approach him. I have often thought, as memory reverted to the old Journal, or as I have re-read the numbers which bear the impress of Mr. Wagner's superior learning, that, though the man is gone, the stamp of his noble character and classical culture is still on these pages, aiding, instructing, elevating, all who are so fortunate as to possess the early volumes of this periodical. (Cook 1884)

Wagner's son, George, would take the helm, albeit temporarily. He kept the journal together for the rest of the year, while searching for a new editor. That position would be taken by Rev. W. F. Clarke, of Guelph, Ontario, then the president of the North American Beekeepers' Society. In his opening words, Clarke declared: "As of old, the *American Bee Journal* will take a straightforward, impartial course, anxious only for the general good." The journal began its new chapter, publishing in Chicago, and is still going strong.

Bee culture can only be regarded as truly "the Poetry of Rural Economy," when it is prosecuted not merely as a source of pecuniary profit, but also as a perennial fountain of intellectual enjoyment. (Wagner 1861)

Washington DC Newspaper Row, 1874

Works Cited

Cook, A. J. (1884). Bee-keepers' Guide: Or, Manual of the Apiary. Lansing, Mich.

Gardner, J. (2012). The Rise and Fall of Early American Magazine Culture. University of Illinois Press.

Langstroth, L. L. (1872). *American Bee Journal*. Edited by George Wagner. Wash. DC.

Pellett, F. C. (1938). History of American beekeeping. Ames, Iowa : Collegiate Press.

Wagner, S. (1861-72). *American Bee Journal*. Edited by Samuel Wagner. York, PA. Wash., DC.

York, G. W. (1903). The story of the *American Bee Journal*. Chicago, IL.

Chapter 8

Ellen Tupper, The Iowa Bee Queen

Ellen Tupper

When I first read about Ellen Tupper, right away I wanted to write about her. It seemed like no one had done a full treatment of her life, her career, and the impact she had on people's lives. Then, this January, a book length article appeared with a long title: Beehives, Booze and Suffragettes: The "Sad Case" of Ellen S. Tupper (1822–1888), the "Bee Woman" and "Iowa Queen Bee." (Mielewczik 2019)

The authors left no stones unturned in their quest to tell everything about this woman's life and times. Much of it had to do with her involvement in gaining equality for women, and fighting the demon rum. I will focus on her connection to the beekeeping world of the 1800s. She went from having a few hives to supplement the family income, to being an editor of the *American Bee Journal*. As the above title suggests, there is much sadness in the story, but inspiration too.

Early Years

Ellen Smith was born in Providence, Rhode Island in 1822. Her family claimed to be descended from famous "founding fathers" of New England, including Captain John Smith and Robert Wheaton on her mother's side. At twelve years old, she and her family relocated to Calais, Maine. Her father was involved with the city's first newspaper as an investor and contributor. Early on, Ellen showed a talent for writing although it got her into trouble when she began writing essays for other students. (Harrison 1870)

In 1843, Ellen married Allen Tupper. Ellen's father Noah Smith and Allen Tupper worked in the lumber and shipping business. The wealthy Tupper family had expected Allen to fall in with their extensive business, but he was leaning towards involvement in ministry, as well as the Temperance Movement and the Women's Rights and Anti-Slavery movements. In 1844 the couple moved to the northern border between Maine and Canada. Ellen had several pregnancies in the ensuing years and lost three of the four children. Only Eliza survived this bout with tragedy, one of many the Tuppers would have to endure.

The Tuppers relocated to Newtonville, a town on the outskirts of Boston, MASS. Ellen recounted many times being told by her doctor, "one of Boston's best," of a heart condition and the likelihood that her "stay on earth would be very short." Of those times in Newtonville, Ellen recalled: "Ah! The weary days and nights of that last year in New England, when nothing seemed to hold me to earth but the clinging hands and loving hearts of my little ones!" (Mielewczik 2019; Tupper 1867)

Like so many people living in the Eastern United States, Ellen was advised to go west to a healthier climate. In one of her many autobiographical essays, she told the story like this:

> I found courage to join my husband in preparations, and before my friends had recovered from the astonishment our crazy plan caused, we were on the way to find a new home beyond the Mississippi. "Gone away from her friends to die among strangers" sighed all who were acquainted with the circumstances. (Tupper 1867)

Out West

The growing Tupper family moved to Brighton, Iowa in 1851 where they purchased 400 acres of land of which some 80 acres were used for farming and raising livestock. Her husband invested in timber land and sawmills. Mrs. Tupper continued:

> In the second summer of our western life came the time that tried men's souls, both East and West. Our little village was not exempt from the scourge. Cholera in its most fatal form visited us and for weeks terror reigned on every hand. When neighbors and friends were hourly summoned; when he, who to- day assisted at the burial of a neighbor, to-morrow, himself filled the plain coffin, hurriedly and without ceremony borne past our door to the grave. When weeping and lamentation were in every house, when any hour might make our remaining children orphans, we could little realize the greatness of our loss. (Tupper 1867)

In the late 1850s, her husband's health became poor and his business began to fail. According to an 1872 newspaper story, the Tuppers had to sell much of their land and their "wealth melted away like dew before the sun." Mrs. Tupper, with children to raise, took a job as a school teacher. Following this same newspaper account, she and the kids all went daily on horseback three miles to the school (Anon. 1872). About this time, Mrs. Tupper decided to get into beekeeping:

> At an expense of twenty dollars I purchased in the Spring of 1859 four hives of bees, of medium strength, and from them secured by fall, fifteen good swarms, and 150 pounds honey in glass boxes, besides some inferior honey. Such honey sold here readily for 15 cents per pound, and each swarm of bees was worth $5, making a gain of $77.50 from an investment of $20. This was an unusually good year. The succeeding year, 1860, was unusually dry, and many bees did not swarm at all, yet mine doubled in number, and I had a quantity of surplus honey. I have no doubt that I can double my number of swarms every year, and realize from 20 to 75 pounds spare honey from each hive besides. (Tupper 1861)

Bees For Women

Ellen Tupper

But beyond the boon of making money for the family, she immediately saw this as something women could take advantage of. Beekeeping in those days was primarily a backyard craft. To be sure, people like Quinby, Harbison, Hetherington, etc. would turn it into an extensive business, but beekeeping has also kept its role as a supplement to income, like chickens or fruit trees. In 1863, The Iowa State Agricultural Society awarded Mrs. Tupper a first prize for her essay on bees. This was an extensive treatise, running over ten thousand words. She clearly connected with the current sources of information. The *American Bee Journal* had just commenced in 1861, but articles about beekeeping had been appearing in the magazines of the times for more than ten years. It is remarkable to me how well informed people were in the mid-1800s.

Information was widely available, and Tupper made clear it should be used:

> The time is now, however, long past when ignorance in the matter
> is excusable, for by the labors of Wagner, Quinby, Langstroth and a
> host of others in our country, information is now disseminated and
> the whole business so simplified that only study, perseverance and
> energy, such as are necessary for success in anything, are needed
> to make this one of the most pleasant and profitable branches of
> rural employment; while the ease with which all parts of the labor
> are performed, peculiarly adapt it to females. Since the invention of
> the sewing machine, many a woman should be emancipated from
> the necessary thralldom of the [sewing] needle which has proved so
> ruinous to the health of the sex. (Tupper 1863)

Editorial Department

Mrs. ELLEN S. TUPPER, Editor.

CORRESPONDING EDITORS:

L. C. WAITE, St. Louis, Missouri.   A. J. POPE, Indianapolis, Indiana.
Mrs THOMAS ATKINSON, Leesburg, Florida.

JAN.   DES MOINES, IOWA.   1874.

Her articles on beekeeping in the *Burlington Hawk Eye* caught the attention of editor William Wilson, and he hired her to write on a regular basis on the topic of beekeeping. On his advice, she was enlisted to write for the "United States Agricultural Report for 1865" of which 180,000 copies were printed and distributed. Then *The Prairie Farmer* magazine hired her as the "special contributor on bee subjects" which she did from 1865 to 1869. Tupper began to exploit the exposure she was receiving through her articles by advertising Italian bees for sale.

> In retrospect, some of the advertisements placed by Ellen S. Tupper
> at that time are, even though very short, quite telling. They show
> for example that she had no inhibitions about making grand claims
> that misled the audience if she thought them helpful. For instance,
> as early as in 1866 she advertised her Italian bee queens as "fully
> tested" and warranted as "pure" even though there was no reliable
> method at the time to do so. (Mielewczik 2019)

Straying From the Path

By the early 1860s, Ellen Tupper was widely known and highly respected as an authority on bee culture. A later writer would look back and declare her "one of the foremost entomologists of the world." She was a proficient writer, skilled at propagating honey bees, and an advocate for the newly introduced Italian variety. However, the question of the purity of Italian bees in the United States took her down a rocky trail. The Italian bee had been imported into the country and various people were selling them at a high price. The only way to tell the Italian from the common honey bee of the time was the fact that the former had orange bands and the latter was black. When the two were crossed the result was variable, and the assumption was that pure crosses should yield golden bees only.

It was known at the time that honey bees mate high in the air, so there were few ways that the crosses could be controlled. One way would be to have the Italian hives isolated; placing them on islands in the Great Lakes was tried. Another plan was to confine the bees until late in the day, when black bees had carried out any mating flights, and then release the Italians to breed – if they were so inclined. The competition was so fierce to advertise pure mated bees, that Mrs. Tupper began to claim that she had achieved what no one else had done: getting the drones and a queen to couple underneath a wire cloth dish cover.

To beekeepers today, who understand bee biology, the idea is preposterous. It took many decades to finally perfect the controlled breeding of bees, using anesthesia and micropipettes. It remains a costly procedure, generally used only for research purposes. But Tupper and some of her supporters maintained that they had been successful using cages of various sorts. How they could have been so misled is a mystery, but eventually the claims died out. The question was discussed endlessly in the journals. One writer averred:

> The discovery was made by Mrs. Ellen S. Tupper, of Brighton, Iowa, who, in a letter to me, dated May 23, 1868, was kind enough to inform me of it, and who then stated that she had made the discovery some time previously. (Anon. 1870)

But after some years passed, clearer heads prevailed:

> *1885 - Cook, Hiller, Heddon, Dadant, Pond, Doolittle, Tinker, and Demaree unanimously agreed that mating in confinement is impossible. (Harbo 1971)*

The Iowa Italian Bee Company

In November 1871, Ellen S. Tupper and Annie Savery together started *The Iowa Italian Bee Company*. They saw beekeeping as a way of empowering women, and went to great lengths to promote it. Annie Savery was born in London, and her husband James Savery, in New York. In 1862, they moved to Des Moines, Iowa where they built up a successful hotel and eventually made a fortune in the real estate business. Annie Savery is best known, however, for her role in the Women's Rights Movement, especially in Iowa.

Savery was an invited speaker at a beekeeping convention in December 1871. She thought she would be speaking to beginners, sharing her new passion with them, but found herself facing a room full of experienced "bee men." Sharp witted, she changed the topic to beekeeping for women:

> *I bought 23 hives of bees, and went to work to learn something about them. It would be uninteresting to you old bee keepers, to state how I proceeded, suffice it to say that I found that every pleasure had its sting. I think I now know the meaning of that phrase, "obtaining knowledge under difficulties." ... As society is now organized, there is nothing for girls outside of marriage, and for this the majority of them are totally unfitted. ... Nothing will contribute so much, and to develop her Into such a woman as every sensible man must admire, as engaging in an employment which will make her his equal. (Savery 1871)*

Advertisements began to appear in all the trade papers for "Pure Italian Bees" sold by Tupper and Savery. Ellen's star was rising. According to Mielewczik:

> *In 1869, she became one of the editors of the "Bee-Keepers' Journal." She thus was, to our best knowledge, the very first woman to ever hold an editorial position in an entomological magazine and perhaps even more generally, any magazine on a biological topic. (Mielewczik 2019)*

In the *National Bee Journal* issue of January 1, 1872, was this passionate description:

> *Mrs. Ellen S. Tupper, also, honored us with her presence. All were glad to see her; every hand was put forth to meet the friendly, cordial grasp of her hand; every one being anxious to have her speak upon her favorite topic—apiculture. To this lady, the bee keepers are under lasting obligations for the numerous instructive articles she has written upon apiculture. Go where you will, Mrs. Tupper's name among bee keepers is a household word. Long may she live. (Anon. 1872)*

Her ability to write, edit, and her renown, propelled her to the position of editor at the *American Bee Journal*, which she held jointly with W. F. Clarke from August 1874 to February 1876. In April of 1875, she was appointed to take charge of Iowa's honey producers exhibit at the celebration of the "One Hundredth Anniversary of the Nation," to be held at Philadelphia, PA, 1876. She began by soliciting contributions from the beekeepers of her state via the *American Bee Journal*. Yet, less than a year later, the *Journal* was to report:

> **The Bee Queen's Temptation**. *Since our last issue, Mrs. Ellen S. Tupper, long known as a writer on bee culture, has "fallen like a star from heaven." On the 28th of January Mrs. Tupper was arrested for forgery. It appears that she has freely used the names of her relatives and friends, and in addition, forged the names of leading citizens of various cities of Iowa, from the name of the governor of the State, down; as well as the names of leading men in the Eastern States. Her forgeries will foot up somewhere from fifteen to twenty thousand dollars, and perhaps more. (Anon. 1876)*

This was followed by comments such as these:

> *"Mrs. Tupper's proverbial philosophy was to forge ahead till she gained $11,000. And now comes emotional insanity with its uplifted umbrella." A prominent bee-keeper in New England, well known to our readers, remarks in a letter of recent date: "I don't wish to say much against Mrs. T–, but if swindling, fraud, and forgery, is any indication of insanity, she has been insane, to my knowledge, for ten years, at least." (Anon. 1876)*

Needless to say, her plans for the Centenary evaporated. She was arrested in 1877. The *American Bee Journal* printed a very short note: "Mrs. Tupper was tried for forgery in Davenport, and upon the plea of insanity, she was acquitted and is now in Dakota on a farm."

Irate customers and creditors showered the journal with requests for money they had sent her for queens which had arrived dead; and for hives, bee equipment and even subscriptions which were never delivered. The publisher, Thomas Newman, was compelled to make it plain that Mrs. Tupper was engaged only as a writer and editor; her debts were her own.

Later Years and Passing

Ellen Tupper disappeared from public view after the debacle. She lived at her husband's farm in the Dakota territory, but he died soon thereafter, in 1879. According to the newspapers, Mrs. Tupper and her daughter Kate moved to Portland, Oregon. She continued writing, contributing to the periodical *Pacific Rural Press* under the pen name "Pioneer." She traveled extensively in California and reported to her friend A. J. King on beekeeping in that state. In 1887, she traveled by ship to Alaska and was there for some time and wrote about it, though nothing remains of her record. Her travels had taken her from the farthest eastern portion of the U.S.A. to the far west. In the end, it was while she was visiting one of her daughters in El Paso, Texas, that she had a sudden heart attack and died, in 1888.

Largely forgotten are her contributions to the advancement of beekeeping and especially bee culture as an occupation to increase the self-reliance of women. Many years would pass before women would rise again to prominence in the field of bee culture. However, the cause of women's rights was taken up by her four daughters. They appeared together as lecturers at the 1894 Woman's Congress in San Francisco. One of the topics was how woman's plight "is much attributable to her clothes. We can't breathe comfortably or sit down comfortably or walk easily. Woman is not physically free."

To be fair, Mrs. Tupper's path was strewn with obstacles and it is amazing that she advanced so far. She overcame ill health; bore 11 children of whom 6 died; she lost to a house fire 200 hives which were being wintered in the cellar. Everyone in the business of shipping bees around the country was negatively affected by the Post Office's outright hostility to the idea of live bees in the mail. Countless

Italian queen bees were lost and had to be replaced due to neglect or deliberate mistreatment by mail handlers. It's hard not to draw a lesson from this story, but in the end my view is Ellen Tupper greatly overextended herself. She aimed very high, perhaps beyond the realm of what was possible to do at the time. And yet, seeing the success of those around her, who would blame her for trying.

Works Cited

Anonymous. (1870). The Illustrated Bee Journal. Vol. 1, No. 7. Indianapolis, Indiana.

Anonymous. (1872). National Bee Journal. Vol. 3, No. 1. Des Moines, Iowa.

Anonymous. (1872). What an Iowa Woman Has Done. Letter from Des Moines. In: St. Louis Globe.

Anonymous. (1876). *American Bee Journal*. Vol. 12, No. 3. Chicago, Illinois.

Cook, A. J. (1875). *American Bee Journal. American Bee Journal*, Vol. 11, No. 4. Cedar Rapids, Iowa.

Harbo, J. R. (1971). Behavioral and physiological aspects of honey bee (Apis mellifera L.) mating. PhD Thesis, Cornell University. Ithaca, NY.

Harrison, E. (1870). Biographical Sketch. The Bee-Keepers' Journal and National Agriculturist 11(1): 1.

Mielewczik, M., Jowett, K., & Moll, J. (2019). Beehives, Booze and Suffragettes: The "Sad Case" of Ellen S. Tupper (1822–1888), the "Bee Woman" and "Iowa Queen Bee." Entomologie Heute [Entomology Today]. 31: 113-227. Aquazoo Löbbecke Museum.

Savery, A. (1872). Transactions of the North American Beekeepers' Society. Cleveland, O., December 6, 1871. Indianapolis Printing and Publishing House.

Tupper, E. S. (1861). Hints on beekeeping. Hints on Bee keeping. 30th March 1861. The Burlington Weekly Hawk-Eye. Burlington, Iowa.

Tupper, E. S. (1863). Essay on bees. in: SHAFFER, J.M. (ed.): Ninth Report of the

Secretary of the State Agricultural Society to the Governor of the State for the Year 1863. F.W. Palmer; Des Moines.

Tupper, E. S. (1867). Why I became a bee-keeper. Letter I. The Prairie Farmer 9(7): 100–101.

Chapter 9

Charles Dadant, A Bee Master's Journey from France

Nearly one hundred years ago, E. F. Phillips, who was then in charge of Apiculture at Cornell University, wrote about "Retaining contact with the past of beekeeping." I want to quote at length from what he said, as it seems to apply to this story:

> At a recent association meeting, I happened casually to say something to one of the beekeepers present about Elisha Gallup. He hesitated a moment and then asked: "Who was Gallup?" He and all other beekeepers should know that Elisha Gallup of Iowa was once a great American leader in beekeeping, the man whom Doolittle among many others looked upon as his teacher.

> What have beekeepers done to show their appreciation of the works of Doolittle, a man who labored late into the night, and night after night, that perplexed beekeepers might have answers to their queries? What bee keepers have shown appreciation for the labors of Alexander, or Charles Dadant, or John S. Harbison who made California beekeeping possible? And this list may be extended almost without end. (Phillips 1926)

Unfortunately, I am afraid my readers are thinking, "Who was Phillips?" Who were any of these people? The Dadant name is familiar to beekeepers throughout the world as one of the principle manufacturers of beekeeping supplies and the publisher of The *American Bee Journal*, for more than 100 years. But I wonder how many know that Charles Dadant, founder of the Dadant & Sons company, lived the first half of his life in France, and came to the United States speaking only French.

Years later, writing competently in his adopted English, Dadant tells the beginnings of "How I became an Apiculturist!"

> I was born in France. My father, a country physician, sent me when six years old to my grandfather, a locksmith, in the city of Langres, for my education. There during nine months in each year, while pursuing my studies, I was between school hours in daily intercourse

with the workmen and learned to handle their tools. And during my vacations—two weeks at Easter, and eight in September and October, I enjoyed country life.

The handling of mechanics' tools was afterwards of great service to me, enabling me to manufacture the various hives which I found described in bee- books, and in treatises on grape and tree culture. Much attention was given to those subjects, and my father's garden was well stocked with trellises and espaliers. Yet, in all the country nothing was so attractive and pleasing to me as the sight of a neighboring hive of bees; so that I sometimes spent hours in watching their labors. (Dadant 1867)

Early Life

Charles Dadant young

Charles Dadant was born in 1817 at Vaux-Sous-Aubigny, a small village of eastern France, the second of seven children. Charles recalls watching the activities of bee hives. The fact that he could see nothing of the inside of the hives only increased his curiosity about them. Knowing of this, the parish pastor invited the boy to look on when he "pruned" the hives in spring. This was the practice of tipping over the straw basket hive and cutting away the tough old darkened honeycombs, in order to give space and encouragement to the bees to build new combs out of fresh white wax.

The pastor dressed young Charles in the customary beekeeper's outfit: a rough linen shirt fitted with an opening to see through covered with heavy wire screen. He never forgot the discomfort and years later he often handled bees wearing ordinary clothes with no extra protection other than a cap over his bald head. Charles was an experimenter and throughout his teen years he built wooden hives after the patterns of Huber and Debeauvoys. These hives were equipped with frames to enclose the honeycombs, but design flaws made them difficult to use. Most French beekeepers stuck with the simple basket hive, and were uninterested in the new models.

Charles' father had expected him to become a country doctor like himself, despite Charles' declaration that he couldn't stand the sight of blood. "That idea will pass," his father assured him, but it did not and in consideration of the costliness of a medical education, the plan was dropped in favor of a career in "business." He was offered a job in a department store in the city of Langres.

Around 1930, Kent Pellett wrote a complete and entertaining biography of Charles Dadant which has never been published. He tells it like this:

> *At eighteen young Dadant returned to the old city and began a life in an atmosphere of lints and woolens. Here he found a certain new air that had been lacking in Vaux and at school: the bustle of trade. From the corners of France merchants dotted the old Roman roads leading to Langres: they came to buy of her locksmiths, her cutlers whose knives were famed over all of Europe, her day goods merchants, and from divers of her tradesmen. (Pellett 1930)*

C. P. Dadant.　　Louis C.　　Maurice M.　　Henry C.　　Chas. Dadant.

Dadant and Sons

In the next several decades, Charles succeeded in business, married his longtime friend Gabrielle, and had three children: a son Camille, and two girls named Mary and Eugenie. The business allowed him to care for his family much as his father had cared for him. During this time he dreamed of an early retirement. According to Kent Pellett: "A secret hope had sprung up within him," to make his living growing wine grapes, with time to tend a garden and bee hives. This dream was dashed when France was thrown into political chaos. Napoleon III was elected France's first president but dissatisfied with that position, he made himself Emperor and started a war against Russia (Crimean War, 1853-56). Meanwhile, at home, the economy collapsed. In Pellett's telling of it:

> Dadant watched the debacle with apprehension. The dry goods firm had over three hundred thousand francs worth of merchandise left from their purchases of that spring. As the panic struck Langres, and the wheels of industry became blocked, the firm saw its business dwindle , and their large stock lost one-third of its value in a few days. (Pellett)

Fortunes seemed to rise again in the late 1850s, as Napoleon III created massive public works projects to modernize France and stimulate the economy. Unfortunately for the Dadants and the people of Langres, being situated on a hill meant the new railway systems bypassed them. Property values plummeted and soon the city streets had grass growing up between the cobblestones. Charles Dadant's story at the age of forty-five was that of "A middle-class urban merchant—failed" (Pellett).

Going to America

> A friend of Charles, one Mr. Morlot, had sent word of cheap but fertile land in Illinois, on which he was making a fortune raising grapes. Sophie, Gabrielle's sister, declared her desire to go with the Dadants to America and offered to pay for everything. Charles left for the United States in April 1863, right in the middle of the U.S. Civil War. He made his way to Morlot's place in Basco, Illinois, where he found life to be vastly different from the settled countryside of France. The land was vast and flat, the houses were rudely made of rough boards, and the roads were deep troughs of mud, navigated by horses and wagons. Morlot also had forty acres outside of Hamilton, Illinois, near the Mississippi River. Charles bought that property and a log house from another French man, and had the logs moved and reassembled on his new "farm." The family arrived in October. Looking back on those times, Charles wrote to one of his friends: "We were reduced to eating brown bread for several years. Neither my wife nor I lost courage, however" (Pellet).

They spent the next few years turning the woods into a farm, replete with a cow, horses, chickens, and a large garden that produced enough that they sent Camille across the river by ferry to sell the surplus in the town of Keokuk, Iowa. He also bought a couple of hives of bees from Mr. Morlot. Despite his lack of knowledge of English, Charles subscribed to the *New York Tribune* and with a pocket dictionary in hand, learned to read it. Once he grasped the language, he bought a book about beekeeping in America. He learned of Langstroth's frame hive, newly invented, which overcame all the faults of both the straw basket hive and the various frame hives that had been tried in Europe. This hive incorporated a gap between the frames and the walls of the box, which the bees kept free and clear as a crawl space. No longer would the frames of the hives get glued tight by the bees and require Herculean efforts to remove them.

Reading in *The American Agriculturist*, Charles learned of Moses Quinby, one of the first commercial beekeepers in the United States. He reportedly harvested twenty thousand pounds of honey in one summer. Incredulous, he asked a friend if such a report was to be believed. His friend said "The American Agriculturist has a hundred thousand circulation because it is never inaccurate." Money was so tight, he could barely afford to buy lumber, so he salvaged boards and crates from town and turned them into beehives. Other local beekeepers like Mr. Morlot still used simple box hives with no frames inside, and resorted to cutting the honeycombs out with a knife. Morlot had one mammoth box which caught Charles' attention. Its owner claimed bees had been living twenty years in that huge box, and despite the eventual rotting of the boards, this bee colony survived while those in the typical smaller boxes failed one after the other.

Modern Beekeeping

Charles Dadant was the first to champion very large hives. Along with this was his use of the centrifugal honey extractor and the newly imported Italian bees. Combining these three factors, he was able to produce large quantities of liquid honey which he began selling in tin pails. Kent Pellett wrote:

> These big hives were adapted for extracted honey. Dadant was producing big crops of it, and selling less of the honey in combs. But the sales of the new product were slow. Customers were still wary at its crystal clearness, where were the cloudy streaks, the stray bits of wax and pollen? Where was the strong smell, the stronger taste of strained honey? They refused to be swindled. And the honey often granulated, which made people more suspicious. Camille had to go

over its whole history each time he tried to make a sale to a grocer, very often only to be ridiculed. (Pellett)

It is worth noting that the most avid importers of Italian bees were people like Samuel Parsons, a successful nurseryman in New York City, and later Charles Dadant, who was skilled in the cultivation of vineyards. Probably they could grasp in an instant how simply introducing the new purebred queen, the colony would soon become "Italianized." In the same way, fruit growers could produce superior fruit by grafting branches from selected trees or vines onto wild but hardy rootstock. Charles began traveling to Italy and by 1872, he had perfected the means of successfully transporting queen bees. According to H. C. Dadant, his grandfather Charles imported as many as 400 Italian queens at a time, with little loss. (Dadant 1950)

Fig. 73.
DADANT HIVE, OPEN.
a, front of the hive ; *b*, slanting board ; *c*, movable block ; *d*, cap ; *e*, straw mat ; *f*, enamel cloth ; *g*, frame with foundation.

Dadant Hive

The last major improvement that brought beekeeping into the Golden Age was the perfection of beeswax comb foundation. Samuel Wagner, editor of the *American Bee Journal*, had secured a patent on the idea, but had never put it to practical use. Briefly, by printing the pattern of honeycomb onto beeswax sheets, and mounting these wax sheets in the wooden frames, bees could be induced to build straight regular combs, which lent themselves to being processed in the modern honey extractor. A. I. Root, the beekeeping pioneer from Ohio, had built the first practical roller presses to make the product. The Dadants saw the value in the sheets but thought that Root was charging too much for them, so they bought one of Root's machines and went into business making their own comb foundation.

In 1882, the honey harvest of Dadant and Son amounted to forty-seven thousand pounds. But not only that: by buying up beeswax from other beekeepers near and far, they had increased their output of comb foundation to twenty-four thousand pounds annually, more than any other manufacturer in the world (Pellett).

Another thing that Charles pioneered was the planting of sweet clover (*Melilotus spp*). To the farmers, it was just another weed brought from the old country by mistake. But Charles saw the value of it as a honey plant, provided that it was widespread. With his usual verve, he set about to propagate it. A hundred years later a very large percentage of the honey sold in the United States came from sweet clover, which spread throughout the midwestern states. Despite the scorn of his neighbors, Charles gave himself credit for its rapid growth along the railroad easements and river banks.

The *American Bee Journal* and *The Hive and Honey Bee*

During these decades, Charles Dadant had made a name for himself as a prolific writer, contributing articles to various publications, including Root's *Gleanings in Bee Culture*, and Samuel Wagner's *American Bee Journal*. When Wagner suddenly died in 1872, there was a rush to find a suitable replacement for him. Charles received the following letter:

> As our old friend, Mr. Wagner has been called to his happy home where we hope he is reaping everlasting Joy, we miss him here below as Editor of our A. B. Journal and I am told you are the only suitable man in America to fill his place as editor. You are better posted in regard to European bee culture than any man I know of

and you are not engaged with any patent hive or bee fixtures and consequently just the man. Can you take charge of the A. B. Journal for us? – H. Nesbit. (Pellett)

True, Charles Dadant had learned to write eloquently in his second language, but he never mastered the spoken word and surprised everyone he met by being unable to express himself clearly, hampered by an impenetrable French accent. But more than that, he didn't want to move to Chicago, where the journal was to be published, so he declined the offer. But his notoriety only increased, and in 1885, he was contacted by Charles Muth, on behalf of Lorenzo Langstroth, who had written the most widely read book about beekeeping, *The Hive and Honey Bee*, published in 1853.

BOOK REVIEW.

Langstroth on the Hive and Honey-Bee. revised, enlarged, and completed by Chas. Dadant & Son. This is the title of the new edition of the Langstroth book, just published by Dadant & Son, at Hamilton, Ills.

The first edition of Mr. Langstroth's work was published in 1852. The last revision was made in 1859, and now after 30 years, during which time more has been done to make bee-culture thoroughly practical than in a century previously, another revision has become necessary, and we are glad to know that it has been done so thoroughly by those eminently practical apiarists, Messrs. Charles Dadant & Son. Former editions have sold rapidly, and so will the present one. It is beautifully printed, embellished with 10 full page plates, and 107 engravings, forming a handsome volume of over 500 pages.

Langstroth felt that in the ensuing thirty years, so many changes had taken place in the beekeeping world, that his book no longer held currency. However, he knew that he did not have the ability to completely revise it, since his wife was gone (she had been indispensable in the earlier versions) and he was perennially plagued by crippling "head trouble," as he called it. Camille Dadant met Muth at a bee convention in Indiana, and from there they traveled to visit Langstroth in Ohio. That October, Langstroth made the trip all the way to the Dadant's place by the Mississippi River.

It almost seems like an unlikely pairing: a mild mannered Minister of the Gospel and a rough and tumble immigrant Frenchman who barely spoke English. But Langstroth had revolutionized beekeeping with his perfection of the frame hive, and Dadant had taken it to the next level by turning the manufacture of beekeeping equipment and supplies into a commercial enterprise. Still, Charles was worried that they may have deep philosophical differences, since he was an avid student of biology and especially the new concept of Evolution. To his delight, he found that Langstroth was entirely open to science and its advancement. But the talk quickly turned to bees and little else was of concern.

They made a contract to work together; Charles had already been compiling material for a book of his own. That winter Langstroth wrote to Charles and said: "I am struggling against the encroachment of that dread disease, and still hope to throw it off. I can only say, go on with your work, and when I am able - if ever I am - I will take hold again." This was followed by a letter from his daughter informing the Dadants that they must continue the work without him. Charles kept on the book, with no word from Langstroth outside of the letters from his daughter. The book was finished and published in 1888, with the title *Langstroth on the Hive and Honey Bee. Revised, enlarged and completed by Chas. Dadant & Son*, published in Hamilton, Illinois by Chas. Dadant & Son. The book has undergone many revisions since then, most recently in 2015 by Joe Graham, and is still published by the Dadant Company of Hamilton.

Dadant factory

Retirement and Beyond

Charles and his dear wife Gabrielle would enjoy long walks in the woods, and when she died in 1895, he wrote to his good friend Eduoard Bertrand that he felt ten years older. In the autumn of 1896, Charles wrote again to Bertand: "You see that we have all in abundance - honey, wine, children." He continued: "Camille is the boss. I do little except to purify the wax, to give a little advice here and there, so that I am now only an old busybody" (Pellett). In one last letter in 1901, Charles congratulated Eduoard on his improved health and remarked that he was losing his own faculties. In 1902, he slipped away, at the age of 85.

George W. York sold the *American Bee Journal* to Charles' son Camille, who moved it to Hamilton in 1912, where it remains. Frank C. Pellett joined the magazine soon afterwards and eventually became an editor. I relied heavily on the work of Frank's son Kent in the telling of this story. Left out were many interesting anecdotes regarding Charles Dadant's politics, philosophy, and theories of beekeeping - worth a book, I am certain.

Dadant House

Works Cited

Dadant, C. L. (1867). How I became an Apiculturist! *American Bee Journal*, Vol. 3, No. 9

Dadant, H. C. (1950). Century of Breeding. *American Bee Journal*, Vol. 90, No. 4

Pellett, K. L. (1930). Charles Dadant: That Bee Man from Champagne. Unpublished manuscript, date approximate.

Phillips, E. F. (1926). Retaining contact with the past of beekeeping. In: Report of The State Apiarist. F. B. Paddock, Ames Iowa.

Chapter 10

A. I. Root and His Early Life

A I Root young

Parsing the history of beekeeping in the United States, one never strays very far from the influence of Amos Ives Root. He is as integral to it as Pasteur is to modern medicine. Were it not for the fact that he lavished his attention on honey production and not something else, like the automobile, he might be as well-known as Henry Ford.

I could feel his excitement when he got involved in beekeeping, especially since he and I both started in our mid-twenties. I remember catching my first swarm and being amazed at how quickly they built the wax combs and filled them with fresh honey. Not very long after young Amos got bees, he began to describe his adventures in the *American Bee Journal*:

About the year 1865, during the month of August, a swarm of bees passed overhead where we were at work; and my fellow-workman, in answer to some of my inquiries respecting their habits, asked what I would give for them. I, not dreaming he could by any means call them down, offered him a dollar, and he started after them. To my astonishment, he, in a short time, returned with them hived in a rough box he had hastily picked up, and, at that moment, I commenced learning my ABC in bee culture.

As a youth, he was early exposed to science (as was I). He was fascinated by electricity, although its development as a source of power and light would come much later. Root made a Leyden Jar, which is an elementary sort of battery which can store a high voltage electric charge. He gave demonstrations of its function at schools, and quite early adopted the dual role as technologist and teacher. In his late teens, he submitted articles to the *Scientific American*. As an old man, he wrote to the magazine, crediting it for his passion for science:

When I was eight years old I was in the chicken business and at the same time experimenting with windmills. Well, the good miller near our home, gave me the sweepings of the mill, if I would sweep it out carefully. In doing this one day I saw a copy of the Scientific American on his table. It took my attention at once and when he saw I was so much interested, he loaned it to me and finally gave me some back numbers and I was so much more interested he finally gave me a big book containing every number. It had then been published about three years. From that time to this I have kept track more or less of what's going on in the scientific and mechanical world through the Scientific American.

The 1800s saw a great proliferation of magazines, eagerly sought after by the reading public, Root among them. It was perhaps inevitable that he would become a major contributor to them, as his readers encouraged his affable and informative style of writing. He would eventually found a publication of his own to give a freer reign to his desire to inform, educate, speculate and also to promote certain moral and ethical ideas. From the beginning, he saw value in quality:

There are a large number of good farmers who refuse to read agricultural papers, because they say, and with considerable reason, that more than half that is written is "impracticable nonsense." We believe the American Agriculturist and the American Bee Journal are at least two noble exceptions. None of their readers can fail to know that each of those papers is edited by some one who is fully posted, and is at home too every time.

Novice Writes for the ABJ

And so, A. I. Root became a regular contributor to the *American Bee Journal* soon after its inception. The publication began in 1861 but was idle during the Civil War, and restarted with vigor in 1865. Root began making monthly contributions writing under the penname of "Novice." He used this persona to endear himself to new beekeepers who justifiably felt themselves to be novices, but it was clear from the onset that he was scarcely unskilled, and soon he began to be peppered with questions. In 1867, he revealed his method: "I had read and re-read the instructions on Italian queen raising until I almost considered myself an expert in the business before trying it." By the end of the same year, his monthly section began with this opening:

> Dear Bee Journal – I do not know how I can better interest your readers in this number than by answering at length a number of queries from a correspondent. He says: "In perusing the Bee Journal I observed your article written on bee-culture, and as I am a bee-keeper, and a sort of novice at that, I thought I would take the liberty of writing to you."

One of the burning questions of the day was how to get pure Italian bees, and how to keep them pure. Root told of his method of multiplying bee colonies, which he calls "artificial swarming," which is essentially dividing existing colonies, almost as you would propagate clumps of irises by setting a portion of the clump away from the "parent." Bees do this naturally when they swarm, but they tend to do it when the beekeeper is not there to recapture them. In any case, he referred to others whose old ideas were receiving punishing blows from the new breed of beekeepers like himself and Charles Dadant. He wrote in regard to the above question:

> One person in particular, a Mr. T. B. Miner, editor of the Rural American, I think quite needlessly exposes his ignorance or something worse, by making the assertion that the Italian bees cannot be kept pure unless on an island or similar place, and that all who claim to the contrary have queens for sale and are cheats and swindlers. … Simply bear in mind that so long as we raise all our queens from one of uniform purity, we can have nothing more impure than hybrids, and very soon a large proportion as pure as the original.

Root Goes into Business

As might be expected from someone whose background included experiments with wind power, Root was an early adopter and promoter of the honey extractor, which had recently been introduced into the US. At this time, most honey was produced by setting boxes on top of the existing hive which the bees dutifully if unwittingly filled with several pounds of honey in each box, which could then be sold as is to the customer. Liquid honey was a bit of a novelty, but with the ease of production, sales soon outpaced those of traditional "comb honey."

HOME APIARY OF THE AUTHORS IN SUMMER.
his photo shows the windbreak of evergreens surrounding the yard. The house-apiary is shown in the background, the upper story of which is used as a workshop. A trellis of grapevines is placed in front of each hive. In summer there is ample shade, and in the fall and early spring the leaves are shed, leaving plenty of sun to strike the hives when it is most needed.

Roots home apiary in the summer

A. I. Root Company

Most beekeepers at the time made their own clumsy extractors out of wood barrels. Root soon began manufacturing lightweight metal honey extractors, but one of his biggest problems was that his customers were using a variety of different sized frames. They were supposed to specify the frame style in use, but often miscommunication led to the wrong sized extractor being sent, and so forth.

A curious fact is that in the early days of hives with frames, often only one box was used. At first, operators would remove the frames right out of this box and put them in the extractor - honey, larvae, pupae and all. They were impressed how the live bees could be spun about with no apparent harm. Root, ever pushing the envelope, realized that by having a second set of frames, honey could be removed and the frames replaced immediately, thus avoiding having to return the original frames. This seems like a small matter to us now, but beekeeping was practiced upon a very rigid line, from the fear that deviation would lead to failure.

The Simplicity Hive

Simplicity Hive

Returning to the subject of different sized frames, the chief reason for the proliferation of varied dimensions was that inventors were keen on patenting their designs. The original Langstroth hive had specific dimensions, and a wide range of "patent hives" were being promoted by self-serving inventors. On the other hand, a hive was considered to consist of one unit, with adequate space inside for the colony to expand in summer.

Charles Dadant, who was an innovator himself, built and advocated a much larger hive after seeing that bees in very large hives seemed to thrive and live longer than those in small hives. So, Dadant's hives had more, larger frames and a much larger box than others. He was keen on extracting honey, so he designed a smaller frame that would fit his honey extractor, and these would be set above the main hive, in what we call a super.

Root, in his effort to make beekeeping affordable, designed what he called the "Simplicity" hive, which he advertised for sale at one dollar. This was a sort of a ploy because it was simply four boards; the frames, the floor and the roof were sold separately. This was the beginning of viewing the hive as a collection of parts, instead of a small fortress. Soon, Root hit upon the idea of stacking these boxes up, allowing the operator to expand or contract the hive as needed. He settled on the original Langstroth frame as the optimal size, but his "hive" consisted of two or more of these to accommodate a large colony. Here he describes the idea, in the *American Bee Journal*:

> *Please, Mr. Editor, can't we have a hive too? We know you will think, and many of our "large family" say, there are too many already and that the more we get, the worse we are off, and that there are patent hives enough for the next thousand years, etc. But, Mr. Editor, "our hive" is "nothing new," and, of course, is not patentable, we hope so, at least, and the novelty, if it is that, it is entirely stripped of the thousand and one valuable, all important features that worry the patient beekeeper and waste his valuable time. Now, then, our hive is simply a square box open top and bottom.*
>
> *When one hive requires more room we simply place as many frames as we wish in another hive with no cover or bottom, and raise the original to allow this to set under it … Accordingly, any number of hives may be piled on each other, or any number of bottoms or covers or all together, and all fit and no projections. They can be packed closely in winter quarters, or in a wagon or in shelter, empty, and if they are to be handled we can walk off briskly with a hive under each arm.*

To modern day beekeepers, this all seems like common sense, but at the time it was quite an innovation. The responses were varied. Some stayed with the single box hive. These are still in use in many European countries. Others opted for smaller frames to create even more modular hives. By and large, the most successful beekeepers were the ones that followed Root's directions, and over time his hive became the standard.

From *Gleanings* to the ABC

Perhaps it was inevitable that so prolific a writer as A. I. Root would launch his own publication in 1873. The initial title was *Novice's Gleanings in Bee Culture*; him still retaining the persona of a novice despite the fact that he was quickly turning bees into a successful business. The *American Bee Journal* seemed to look upon Root's new venture with favor:

> *Our valued friend and contributor, "Novice," desirous of even more than that unrestrained liberty which he enjoys in these columns, has started an unpretentious little quarterly at twenty-five cents*

a year, which he entitles, "Novice's Gleanings in Bee Culture, or How to Realize the Most Money with the Smallest Expenditure of Capital and Labor in the Care of Bees, Rationally Considered." We offered to husband his "Gleanings" for him in the Journal, but he prefers to keep them in his own hand. And, provided always that Novice does not stint his communications to the Journal, we can have no objection to his commencing the "Gleanings," or any other periodical, and we wish him the fullest success.

Root wrote extensively about the vexation his new venture brought him. At first he worried that he would fail, but after the first issue he decided to go from a quarterly to printing monthly. One of the things that irked him was the frequency of typesetting errors, which often changed the whole meaning of a sentence. To try to present his writing more clearly (to the printer) he purchased a typewriter, which in the 1870s cost $125, at a time when the magazine was selling for 75 cents a year.

By far the most frequent complaints were about the way the material was presented. Some readers chided that the articles were too superficial while others lamented that they were too long and included "so many repetitions of things that everybody knew already." At some point he conceived the idea of an encyclopedia, called "the ABC," that would contain everything that was currently known about beekeeping, placed alphabetically so that the reader could go directly to the material on whatever question was burning him up at the time.

He told the readers that he invested a small fortune in printing type (letters). Normally, type was set by hand and after an issue of a paper or magazine was printed, the metal types were returned to their "type-cases" to be used in the composition of the next issue. So, to have the types of each page of a book kept assembled, required a much larger investment of these thousands of individual letters. He explained it to his readers:

I do not wish to write a single thing for the ABC without at least some practical experience in the matter. I am selling a great many books on bee culture, and many are the questions asked in regard to their teachings. I can not be responsible for the teachings of other writers, but I do intend to be responsible for all that appears in the ABC; and furthermore, I have been to the expense of purchasing the type for the whole of it, that every mistake or wrong statement may be corrected just as soon as it is found to be such. The sheets are to be printed only as fast as they are sold, that none of the information may be old or behind the times.

Root printed the first eight pages of his "ABC Book" in the May 1877 issue of *Gleanings*. As promised, it was an alphabetical encyclopedia format, commencing with an entry on the Age of Bees, as it had already been observed that bees could survive for six months of winter, but in summer often lived but six weeks, during the harvesting period, worn out from the arduous work of flying to and from the clover fields. This was followed by Alsike Clover and an entry on the Apiary. While many beekeepers then as now, have bee yards that resemble junk yards, Root was very keen on having the apiary aesthetically pleasing. He promoted the idea of planting a grape vine close to each hive, not only to please the eye but to provide shade in summer. These days plants in the bee yard are considered annoying or a fire hazard, depending on where you live and how many hives you have.

The ABC of Bee Culture as a complete book, came out in 1879. It has been in print ever since and has gone through many revisions, each enlarging and expanding upon the ideals of the original. I would say that no bee book collection would be complete without Root's *ABC & XYZ of Bee Culture* and the Dadant's latest update of *The Hive and the Honey Bee*. Personally, I have six different editions of the ABC and several of the H & HB. By the way, all of these periodicals and books have been digitized and are freely available where the copyright has expired. Of course, it is still essential to keep up-to-date on the latest news and information, via monthly publications.

BEES AND HONEY.

Our 12th Edition
Illustrated Circular & Price List,
OF
Implements for Bee Culture with Directions for their Use.

A. I. ROOT, MEDINA, O., JANUARY 1st, 1878.

Implements for the Apiary.

No. 1, shows a Simplicity Hive, single story, with the sheet of Duck removed, so as to show the 10 frames in place. The Chaff Cushion is shown in the cover, where it is fastened by 8 or 10 tacks around the edge. You will observe that when the Cushion is thus fastened in the cover, we are obliged to have the sheet of Duck shown at No. 8, fitted closely over the frames that the bees may not get to the Cushion, or it would be stuck so tightly to the frames that we could never get the hive open. For wintering, a much thicker cushion is used, unquilted, and placed in an upper story. This Hive is shown with the entrance closed, by pushing it back squarely on the bottom board, while Nos. 2 and 3 are pushed forward so as to give a ⅜ inch passage for the bees. No.

A. I. Root catalog 1878

Perfection of Comb Foundation

Printing the honeycomb pattern onto beeswax sheets, which would then be mounted in the wooden frames in order to get regular uniform honeycombs in the frames, was dreamt of early on, and even patented before it could be brought to practical use. Many people tried putting wax into presses like waffle irons, but the resulting product was often too brittle to be of use. We have already seen that Root was not just an avid inventor, fascinated by electricity and mechanical engineering, but also a businessman with a money-making attitude.

In 1876, Root painstakingly described the geometry of the bee cell, which has invoked admiration in people from the ancients to the modern times. He referred to the geometry of the *rhombic dodecahedron* and gave detailed directions on how to cut metal types at the correct angles to produce an accurate replicate of natural honeycomb. Root conceived of printing foundation in a manner like printing flyers, with a hand cranked printing press, and soon offered these for sale as well as the mass produced beeswax foundation sheets. Most beekeepers found it less vexing to buy them than to try to make the sheets themselves.

A GLASS PAPER-WEIGHT SHOWING THE MATHEMATICS OF THE HONEY-COMB.

A TRUE DODECAHEDRON.

We have finally succeeded in getting a thousand glass paper-weights, such as are described in our heading. Before we could get them, the order had to be sent to Germany to have them made, and it has taken a little over a year and a half to get them out. The adjoining cut, taken from the A B C book, shows what they are like.

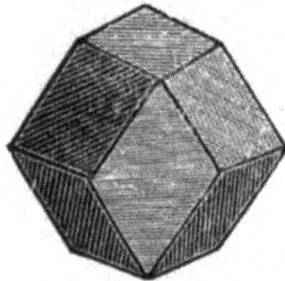

A TRUE DODECAHEDRON.

Dodecahedron

I spent the better part of the 1970s running a foundation mill myself, at Henry Knorr's Beeswax Factory. I had no idea at the time how large a part Root played in the perfection of the process. The most significant innovation was the sheeting machine which could produce a continuous ribbon of beeswax of any width or thickness. This continuous output would be automatically rolled up into individual rolls and stacked to cool. Then the ribbon of wax would be fed into the rollers of the foundation mill, now powered by a variable speed electric motor. The output would be cut into individual lengths, by a chopping blade, and then stacked by hand into neat piles, to be weighed and placed in cartons for shipping.

My output was usually about one thousand pounds a day, depending on interruptions. I was also expected to mind the bee supply store. This involved selling hives and equipment but also receiving and weighing large lots of bulk beeswax. The factory also produced candles using the same sheeting machines, equipped with round openings in place of the slot. Out of these openings, continuous candles would emerge, which initially would be cut into six foot lengths and resized later when cool. One of Henry Knorr's lifelong challenges was to figure out how to include the wick in this stage of the process. Ultimately, the wicks had to be threaded into the candle by hand, as they always had been. Both Henry and his father, Ferdinand, lived into their late 90s; Henry passed away in 2018. I hope to recount their story some day.

Novice's Honey Extractor, A. I. Root Co., Medina, Ohio. Honey is thrown from the decapped cells by centrifugal force.

Novice's Honey Extractor, A. I. Root Co. Medina, Ohio, mounted on an old cream separator base, was bought secondhand by the late Samuel Berger of Shartlesville in 1894.

Novice's extractor

Parting Words

A I Root

Root was a pioneer in every sense of the word, an inventor, a promoter of new ideas, a maverick. I would like to close with an anecdote, told by Root's longtime friend C. C. Miller. I will explain the reference to milk wagons, for those who don't know. At one time milk was delivered to houses in the wee hours of the morning, to be there to greet us at breakfast time. This practice continued with trucks well into the 1960s.

"Over forty years ago," said Dr. Miller as he settled himself in the pillows of the bed, "that man A. I. Root slept in the same bed with me, and kept me awake until midnight telling me how he was going to make a fortune at bee raising. He had a scheme to tap the maple-trees, and run the sap direct into the bee- hives and supply honey with a maple flavor. Last night, forty years later, he told me of another scheme until I had counted thirteen passing milk wagons."

Works Cited

American Bee Journal (1861 to present)

Gleanings in Bee Culture (1873 to present)

Scientific American (1845 to present)

Chapter 11

The Beginnings of Scientific Apiculture

The nineteenth century must always stand out in the history of the world as the period which has combined the greatest development in all departments of science with the most extraordinary industrial progress. It was not until this century that scientific investigation used to their full extent the twin methods of observation and experiment. (Schuchert 1918)

Honey Bees and Science

Beekeeping today falls into three main categories. Commercial beekeeping involves thousands of hives for the production of honey or pollination of crops; apiculture research is conducted mainly at University and Government Labs. The third class has by far the most people: those who have several or up to a hundred hives, primarily for the experience of working with bees on a scale that suits them. In this article, I will explore the history of scientific apiculture in 19th century America, and how it affected beekeeping as a whole.

The word science derives from the Latin *scientia*: knowledge (knowledge as opposed to belief). However, the concept of a *scientist* came about fairly recently. In 1840, William Whewell wrote in his treatise about science and its history:

> We need very much a name to describe a cultivator of science
> in general. I should incline to call him a Scientist. Thus we might
> say, that as an Artist is a Musician, Painter, or Poet, a Scientist is a
> Mathematician, Physicist, or Naturalist. (Whewell 1840)

In the United States of the 1850s, there was a great outpouring of books about the honey bee. Up till this period, bees were seldom considered as anything more complicated than, say, chickens, and many people had a few hives. At the same time, inquisitive individuals yearned to know more about the inner workings of the bee colony. One of the books of this era was K. P. Kidder's *Guide to Apiarian Science*. He gives due credit to the French pathfinder François Huber, who despite being blind, made more discoveries than most of his contemporaries, aided by the keen eyes of his assistants. Kidder wrote:

> Apiarians owe more to Huber, for the advancement of Apiarian

science, than to any other man. Having in many instances verified some of the most important of his observations, I take the greatest pleasure in acknowledging my many obligations to him, and holding him up before the world as the founder of Apiarian science in a great degree. There are few men, either in ancient or modern times, that have had such means, perseverance, and patience in carrying on his many costly experiments for a series of years, as the celebrated Huber. (Kidder 1858)

One of the tools that facilitated Huber's discoveries was a hive with the combs mounted in moveable frames. These were attached with hinges so the hive opened up like a big book and its inner workings could be examined at will. However, the design was cumbersome and far more difficult to manipulate than the frame hives that were popularized in the United States in the 1860s.

In the "Annual Report of the State Board of Agriculture" of Missouri for 1870, T. R. Allen of Allentown sums the story nicely:

Quinby's "Mysteries of Bee Keeping Explained," appeared in 1853. This valuable work on the subject gave a new impulse to bee culture in this country to an extent previously unknown. It has passed through several editions, and is still a standard work on apiarian science. "The Hive and Honey Bee," by L. L. Langstroth (1853) is also a most valuable standard work. To this author we are also indebted for that most inestimable invention, the movable comb frames, making another great epoch in the science of bee culture. (Allen 1870)

What is notable in this excerpt is the deliberate use of the term science in connection with the keeping of bees. It was understood that if the practice was to advance measurably, a clear and deep understanding of the natural history of honey bee colonies would be required.

The *Scientific American*

The people of the United States clearly had an appetite for information about scientific studies: the *Scientific American* began as a four-page weekly newspaper in 1845, which makes it the oldest continuously published magazine in the US.

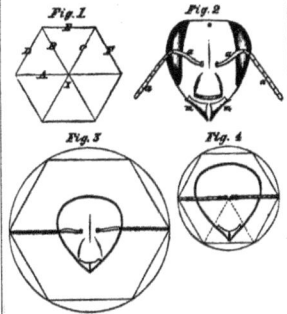

THE HEXAGONAL CELL OF THE HONEY-BEE.
BY W. J. WEEKS.

During this time period, the magazine was promoting the patenting of inventions and its pages were littered with various bee hive modifications, but it also presented the scientific side. In 1860, an article appeared describing the means by which honey bees are able to accurately measure and construct their wax comb cells, the beauty and regularity of which for centuries has astonished the human mind. W. J. Weeks informed the readers that they use their antennae to measure the width of the cells. He wrote:

The intimate relation between the length of the antenna; and the size of the cells was discovered by the author of this article, in the year 1852, he being previously acquainted with the properties of the hexagon. Any one, knowing this relation, may now understand how thousands of cells in a single hive may be all of one form and size, how every individual cell of these thousands may be precisely similar to every cell of the aggregate millions in all other hives, and how, the world over, wherever this species of bee (Apis mellifica) exists, all its regular hexagonal cells of the two classes–worker and drone –can be exactly equal each to each, for every adult worker in its antenna; is provided with an equal rule and compass.
(Weeks 1860)

Eventually, this famed regularity was shown to be rather variable, but only when subjected to very close measurements. On the whole, the comb pattern is remarkably uniform, and so it is also with wasps that make equally regular cells out of paper they form using plant fibers. In any case, the attention being paid to the natural history of honey bees was being presented to a wider audience.

The following year the *Scientific American* described the problems regarding the effort to export bees from the East Coast, to the West:

Notwithstanding disastrous results attending the previous years' shipments, there were upwards of six thousand hives of bees imported during the winter of 1859-60. They arrived in better condition apparently than those of the previous year ; yet, owing to the fact that large numbers of them were infected with the disease known as foul brood prior to their purchase and shipment, together

117

> *with the effects of so long a voyage, probably one-half of the whole number were lost. Many of the remainder have since died, or now linger in a diseased condition, which is infinitely worse for the parties owning them than if all had died at once. Thus, the result was bad for all concerned : for, while some have lost their money, others have injured their reputation, besides paralyzing for a time an important branch of productive industry. (Anon 1861)*

This is notable as it foreshadows one of the major problems the bee industry would suffer, communicable disease, which was not really brought into control until the widespread use of antibiotics, nearly one hundred years later. In fact in the 1860s, at the time of this report, the understanding of bacterial infection was just coming to light through the discoveries of Louis Pasteur, in France. Further, Pasteur would identify the microorganism *Nosema bombycis* in the silkworm, which presaged the discovery of the related species *Nosema apis*, in the honey bee. This, too, would come to be treated with an antibiotic in the 20th Century.

K. P. Kidder

KIDDER'S GUIDE

TO

APIARIAN SCIENCE

Price 50 CENTS.—Sent by Mail 57 Cents, or 19 Three Cent Postage Stamps.

K. P. Kidder

The author of *The Guide to Apiarian Science*, K. P. Kidder prefaced his book with a biography of Huber whom he referred to as "The Prince of Apiarians." The book is arranged much as a textbook would be today. The first three chapters cover the topics of anatomy, senses of the bee, and impregnation of the queen, respectively. While the exact mechanism of the fertilization of eggs had not been determined, Kidder rightly suggests that the colony benefits from out-crossing the queen with drones from another hive and as he says, "still better from a neighboring Apiary." As we know now, the crossing of very closely related queens and drones can result in the queen being unable to generate female offspring due to homozygosity at the sex locus. Again, another hundred years would have to pass before this would be understood.

Kidder leaves the more practical information on beekeeping technique till the later chapters: Chapter 16 is about hives, and 17 covers "General Management of Bees." This is followed by a complete index of the book, making it readily accessible. Beekeepers in the mid 1800s were well supplied with textbooks. A little bit about the man himself:

> Kimball P. Kidder was born on the family homestead at Randolph, Vermont, November 10, 1821. A man of brilliant parts, his ingenuity and inventive skill brought him into National prominence, two of his patented devices (of utility to the bee farmer) gaining him much credit in their extensive adoption throughout the country. His abilities and integrity brought him into the high regard of the people of Burlington, Vermont, in which city he made his home during the latter part of his life. He had passed his seventy-first year when, in July, 1893, his life came to an honorable end, a life which had been full of meritorious effort and outstanding ability. (American Historical Society 2000)

A. J. Cook

A. J. Cook

Albert J. Cook was born Aug. 30. 1842, at Owosso, Michigan. He entered the Agricultural College at 15, but illness intervened. He graduated by age 20 and went to California with the hopes of improving his health. Cook then studied with Agassiz at Harvard University and traveled full circle, back to the Michigan Agricultural College. He worked as an instructor and then in 1868 he was made a Professor of Entomology. A. J. Cook taught the first course in beekeeping in an American agricultural college, in 1871. In Cook's words:

> I once heard a well known professor and scientist, than whom there is no better student of American agriculture, remark, that the art of agriculture was founded almost wholly upon empiricism; and that all it had to thank science for, was that the latter explained what had already been determined by the empiric method. Whether this be true or not, the reverse is most certainly true of practical entomology. Economic entomology rests almost wholly upon science. So, too, apiculture, as practiced to-day, owes its very existence to science. (Cook 1881).

Cook should be remembered for *The Bee-keepers' Guide - or - Manual of the Apiary*, which first appeared in 1875. Here are some of the reviews:

> *It is a book which does credit to our calling; one that every bee-keeper may welcome as a fit exponent of the science which gives pleasure to all who are engaged in it. - American Bee Journal.*

> *It is just what might have been expected from the distinguished author - a work acceptable to the ordinary bee man, and a delight to the student of scientific apiculture. - Bee-Keepers' Magazine.*

> *"Manual of the Apiary" contains, besides the description of the anatomy and physiology of the honey-bee, beautifully illustrated, the products and races of the bees, honey plants, the instructions for the different operations performed in the hives. All agree that it is the work of a master, and is of real value. - L'Apiculteur, Paris.*

> *It is the fullest, most practical and most satisfactory treatise on the subject now before the public. - Country Gentleman.*

> *I have derived more practical knowledge from Prof. Cook's new Manual of the Apiary than from any other book. - E. H. Wyncoop.*

In 1894, Prof. Cook decided to leave Michigan, and go back to California, where he obtained a position as the chair of Zoology at Pomona College. In his seventies, he returned to Owosso, Michigan and died there, in his birth town.

C. V. Riley

Closely paralleling Cook's life, was that of Charles Valentine Riley. He was born in 1842 in London, England and educated in France and Germany. During that period he distinguished himself by collecting insects and producing prize-winning drawings and paintings of this subject of interest to him. When he was 17, he moved to a farm in Illinois, about 50 miles outside of Chicago. By 21, he was working as an artist, reporter, and editor of a leading agricultural magazine, *The Prairie Farmer*. He died about age 52, in a bicycling accident, cutting short

Riley at his microscope

an abundant life.

C. V. Riley was named Missouri's first State Entomologist in 1868. As Riley's career advanced he became the Chief of the U. S. Entomological Commission in 1876, and Chief Entomologist for the U.S. Department of Agriculture in 1878. Riley brought about one of the first success stories in the biological control of pests. In 1888 he introduced the Australian vedalia beetle to combat scale insects, thereby helping to save the California citrus industry. Further, Riley figured in research which helped rescue the French wine industry from another insect pest. The USDA has created a well-supplied website about him. Here is a quote:

> Riley was truly a "whole-picture" person - an artist, a poet, a writer, a journalist, a linguist, a naturalist, and a philosopher as well as a scientist. He also added cultural aspects to his work through his charter membership in the Cosmos Club in Washington, D.C. In 1878, Riley joined with Alexander Graham Bell, John Wesley Powell, and fifty-seven other men in science, literature, and arts to found the Cosmos Club as a "social club for individuals of distinction and sociability." (USDA 2021)

Riley's connection with beekeeping can best be illustrated by this excerpt from "What the Department of Agriculture Has Done, and Can Do, for Apiculture," which appeared in the *American Bee Journal*, February 16, 1893:

> Considerable has been done by the Department, and through its agency for bee-keepers – much more, probably, than most of you are aware of – the published reports of the Department show. These reports, hundreds of thousands of which have been distributed very generally over the land, have surely had their influence in the promulgation of intelligent and humane methods in the culture of bees. Beginning about the time of the first edition of Langstroth's celebrated work, or nearly a decade before any bee-periodical had been printed in the English language, the Department reports have from year to year given some notice of progress in bee-culture, statistics of honey and wax production, and on several occasions excellent little treatises on bees and bee-management. (Riley 1893)

Dr. Riley went on to delineate the primary areas he wished to see investigated scientifically. These were, in brief: 1. the introduction and domestication of races of bees reported to possess desirable traits and characteristics; 2. to make experiments in the crossing and mingling of races already introduced, and such as may hereafter be imported; 3. to make experiments in the methods of artificial fertilization of queen bees; 4. to study the true causes of diseases yet imperfectly understood; and, 5. to obtain incontestable results by intelligent experiments on scientific methods, as to the capacity of bees to injure fruit (still hotly debated at the time).

Frank Benton

Frank Benton

Frank Benton was born in 1852, in Coldwater, Michigan. According to his son, Ralph, Frank "very early became interested in the study of bees. He learned the printer's trade as an apprentice in his father's printing office, and at 15 went to sea, sailing before the mast for two years, rising to the position of third mate of a barkentine on the Great Lakes" (Benton 1908).

Benton graduated from Michigan Agricultural College with his B.S. in 1879. He also studied at the University of Tennessee, the University of Munich, and the University of Athens. In 1879, he taught modern languages at Michigan. Around 1880, he was the investigator in charge of apiculture with the Bureau of Entomology in the United States Department of Agriculture. As a student, Frank came under the influence of Prof. A. J. Cook, and continued his association with him through the years.

Frank Benton should be best remembered for his travels to find and import various sorts of honey bees from across the globe. According to his son, in 1879 Frank went - together with his new wife Hattie and D. A. Jones, an ambitious beekeeper from Ontario, Canada - on a voyage which took them to regions as widely flung as Europe and Asia. After a year, Jones returned home but Benton stayed on. He went from Palestine, to India, Singapore and Indonesia. Ultimately he and his family settled for a time in Munich, Germany where he managed the "Bavarian Apiary," established by him as a shipping center for the various sorts of bees he had collected. Again, according to his son Ralph Benton:

In 1885, upon the presentation of a dissertation written in Germany, Mr. Benton was honored with the degree of M.S. from his alma mater, Michigan Agricultural College. In 1886 the Bavarian Apiary was discontinued, and the Carniolan Apiaries established, first at Laibach [Ljubljana], and two years later at Krainburg [Kranj], in the Province of Carniola, in southern Austria [present day Slovenia]. This resulted in the study and introduction of that most valuable variety of bees, the Carniolans. (Benton 1908)

Benton's exploits were chronicled in the *American Bee Journal*. This is from a 1919 article entitled "Benton's Travels":

During the years spent in Munich several trips were made to Cyprus and Syria, and on one occasion Tunis and the African coast were visited and the bees of these regions studied. Italy was visited by the way as was also the little province of Carniola, in southern Austria, with the result that the four years from 1886-90 were spent in the fastnesses of the Carnic Alps in investigating, breeding and giving to the world the docile bees native to these mountains. (Anon 1919)

FIG. 57.—Benton queen cage. This is the cage usually used for sending queens by mail

Benton Queen Cage, 1885

When he finally returned to the U. S., he was faced with a difficult choice.

Dr. Benton was offered a chair in modern languages at Cornell University, and at the same time came an offer from the United States Government to go into scientific work at Washington. It was not an easy matter to decide, especially for one so rarely gifted in both fields of endeavor. But at the parting of the ways Dr. Benton, at the age of 39 years elected to go into scientific work, thereafter becoming only indirectly identified with academic life as an occasional lecturer. He proceeded to Washington in July, 1891. (Anon 1919)

By 1891, C. V. Riley was head of the USDA Entomology Division, and both A. J. Cook and Frank Benton were listed by him as Michigan field agents, so these three men were closely linked throughout their lives. Frank Benton's book, *The Honey Bee: A Manual of Instruction in Apiculture,* was published in 1895, under the auspices of the USDA. It is a fitting culmination of the 19th century, which brought scientific apiculture out of the dark into a new era.

Works Cited

Allen, T. R. (1870). "Bee Culture." Annual Report of the State Board of Agriculture. Jefferson City, Missouri.

American Historical Society. (2000). Tercentenary of New England families, 1620-1922. Salem, Mass.: Higginson Book Co.

Anon. (1861). "Exportation of Bees to California." Scientific American, Vol. 5, No. 8. Anon. (1919). "Benton's Travels." *American Bee Journal*. Vol. 59, No. 12.

Benton, Ralph. (1908). "Government Help in Apiculture." Pacific Rural Press. San Fran.cisco, California. Vol. 76, No. 22.

Cook, A. J. (1881). "The relation of apiculture to science." The American Naturalist. Vol. 15, No. 3.

Kidder, K. P. (1858). Kidder's Guide to Apiarian Science: Being a Practical Treatise, in Every Department of Bee Culture and Bee Management. SB Nicholas.

Riley, C. V. (1893). "What the Department of Agriculture Has Done, and Can Do, for Apiculture." *American Bee Journal*. Vol. 31, No. 7.

Schuchert, C., & Dana, E. S. (1918). A Century of Science in America: With Special Reference to the American Journal of Science, 1818-1918. Yale University Press.

USDA. (2021). Charles Valentine Riley Collec.tion. National Agricultural Library Special Collections. https://specialcollections.nal.usda.gov

Weeks, W. (1860). "The hexagonal cell of the honey-bee." Scientific American. Vol.

2, No. 20. Whewell, W. (1840). The philosophy of the inductive sciences: founded upon their history. JW Parker.

Chapter 12

Large Scale Beekeeping:

How It Came About in the U.S.

Life is too short, and Americans are too busy to spend the time necessary to delve into a multitude of volumes in order to post themselves on the general history of the past. –
Thomas Newman, 1886.

Scaling Up

Large scale beekeeping probably started in Egypt, where hundreds of traditional cylindrical hives made from clay can still be seen stacked together making long, high walls of hives. *The Beekeepers' Record*, a British periodical from 1882, mentions "very large numbers of hives owned by single individuals (from one hundred to two thousand)." In the United States, prior to the 1860s, beekeeping was generally done on a small, part-time scale.

The management of large numbers of hives, whether they were made of crockery, or basket-like skeps of Europe, was fairly simple, so it could readily be scaled up. The modern wood frame hive, implemented in the 1860s, facilitated complex manipulations but was not immediately scaled up, being so labor-intensive. However, beekeeping pioneers of that era, like John Harbison and Moses Quinby, showed that beekeeping could be carried on a large scale with the new hives. Having hundreds or thousands of hives built on a standardized pattern simplified many operations.

Even today, beekeeping can be divided into two distinct categories: intensive and extensive. The intensive plan involves very close monitoring of the internal workings of the honey bee colony, such as replacing old queens with young ones, and other skillful procedures. At the opposite end is beekeeping which relies on very large numbers of hives to produce huge amounts of honey. By paying far less attention to individuals, the unit is the apiary rather than the hive. Systematic management allowed beekeepers to harvest the honey from one hundred hives in the same amount of time a traditional beekeeper might manage to do ten.

Another factor in the move towards large scale beekeeping was the mass production of hive parts from economical lumber. Both Harbison and Quinby had significant experience in cabinet making and working in sawmills, so they were able to quickly conceive of the technique and benefit of making hundreds of hives on the same identical plan. Ironically, neither of these two individuals adopted the Langstroth hive, which eventually became the standard. Harbison's hive was like a cabinet which opened from the side, instead of the top like the Langstroth hive. Hives on this plan were popular in Europe and can still be seen in some areas there. John Harbison went to California during the Gold Rush and probably would have made a fortune no matter what field he went into. Initially, he tried his hand at fruit growing, which as we know, became one of California's greatest strengths. He got involved with the lumber business, which was harvesting immense quantities of old growth redwood trees, the irreplaceable product of millions of years of thriving in the unique microclimate of the Northern California coast. This lumber is light weight, rot resistant and easy to mill, so it was used for everything from building houses, hives, even water pipelines and huge wine vats. But Harbison saw the future of beekeeping in California and set about to import honey bees to where there were none. Several publications had sprung up, ready to tout the burgeoning fortunes of honey producers. A. J. King wrote at the time:

> Mr. J. S. Harbison reached New York, in 1876, with his great shipment of honey, produced in his six apiaries in San Diego county, California. This shipment consisted of ten [freight train] car loads, each containing 20,000 pounds, or 200,000 in all. In Los Angelos [sic] county we are told that there are not less than 200 apiaries, and over 12,000 hives, from which over 500,000 pounds of surplus honey are taken annually. The income of Mr. J. S. Harbison, derived from honey alone, is said to be more than $25,000 per annum, over and above all expenses. In the State of New York, Captain J. E. Hetherington, of Cherry Valley, sold, in 1874, over 58,000 pounds of honey from his own apiaries, and Adam Grim, of Jefferson, Wisconsin as much more. But we will not further name individual incomes. (King 1879)

New York State

Meanwhile, back in the East, a number of people such as Quinby, Root, and others were involved in the development of beekeeping as an industry. The

American Bee Journal included among these, a lesser known but equally successful beekeeper, Capt. Hetherington:

> "While California is the greatest bee-ranching or honey region in the world, owing to the excellence of its climate and the endless variety of its honey-yielding flowers, the quality of its honey does not excel, even if it equals, that of the honey produced in New York State," said a large wholesale dealer.

> "One of the most extensive bee culturalists in this or any other country, is Capt. Hetherington, whose apiaries along the Cherry Valley Creek, in Schoharie county, annually turn out over 100,000 pounds of the choicest honey. It takes nine men and two steam saw mills five weeks to prepare the lumber for the boxes in which the honey is made by his bees. Nearly 150,000 panes of glass about six inches square, are used in these boxes.

> "Capt. Hetherington has at work, this season, nearly 2,500 colonies of bees. These are not all on his own premises but are scattered among the orchards and fields of farmers along the creek to whom he pays a rent for the privilege of his bees working in the clover, buckwheat, or whatever blossoms are in season on the farm. The care of these bees does not fall upon the owner of the land. Capt. Hetherington keeps men and teams constantly employed looking out for them. He has received as much as $26,000 for one season's crop." (Anon 1884)

Capt. Hetherington

It would be a mistake to leave out the significance that consumer demand had for this bonanza. Downstate in the New York City metropolitan area, prosperity drove the market for high quality agricultural products. Natural honey comb, which the beekeeper had induced the bees to build in small wooden boxes (with the glass sides added, post-harvest) was eagerly sought after, whether it was from California or Upstate NY. The invention of the honey extractor led to the acceptance of honey sold in cans, which greatly expanded the usefulness of the product and the demand for it. At the same time, honey production was double or more using the honey extractor, because the honey combs could be emptied and refilled over and again. Some beekeepers, like Harbison, never adopted machine extracting, preferring to continue to offer the fancy honey comb. That is, until he got out of beekeeping and into the San Diego real estate business.

Moses Quinby and Capt. John Hetherington lived and worked close to each other. The former lived in St. Johnsville, by the Mohawk River and the latter in Cherry Valley, some twenty miles to the south. Quinby had learned through the writings of T. B. Miner that the prospects for beekeeping were much better further Upstate and he moved there from his home in Coxsackie, NY in 1853. Hetherington met him soon thereafter, when Quinby was still a teenager. When the American Civil War broke out, Hetherington enlisted; his exploits in the war have been written about elsewhere (Edwards 2014). In short, he was made a captain and was henceforth referred to as Capt. Hetherington.

The Coggshalls of West Groton

William Lamar Coggshall

The Coggshall brothers, David H. and William Lamar were born and grew up in Groton, NY; the former in 1847 and the latter in 1852. Groton is about 15 miles north of Ithaca, which is the home of Cornell University, located on the hills above Cayuga Lake, one of the largest of the Finger Lakes. At the time, the region looked nothing like it does today. Most of the vast forests of the East had been cut down; the Finger Lakes Region was no exception. It had been transformed into a great farming region with access to the New York markets via canals and later the railroads. Among the crops were white clover (*Trifolium repens*) and buckwheat (*Fagopyrum esculentum*), both phenomenal nectar plants. Ironically, clover honey is very mild and fragrant with the scent of the blossoms whereas buckwheat honey is dark, pungent and often not appreciated by people unaccustomed to it. Yet, customers existed for both types. The region also had (and still has) large numbers of basswood trees (*Tilia americana*), a prolific albeit unreliable nectar source. By mid 20th century, clover and buckwheat ceased to be common but the forests began to grow back in many areas, especially where farming was never really that good. Corn is the predominant field crop these days.

I was introduced to the Coggshalls' story by the photographs of Verne Morton, who – using a large format camera with glass plate negatives – documented country life in the vicinity of his home in Groton, during the late 1800s and early 1900s. His negatives are conserved by The History Center in Tompkins County. These show very clearly many of the rural activities at the time, and in particular Verne took a lot of photographs of the Coggshall brothers' homes, bee hives and operations. These photos as well as the stories that accompanied them appeared frequently in the beekeeping magazines of the times, and scarcely a beekeeper in the US and abroad had not heard of the name Coggshall. What follows is a first-hand account by Lamar Coggshall. I must mention that they always had a good sized shed at each apiary, here called an out-yard.

> *My first out-yard was establisht in 1878. I have now three, ranging from three to 26 miles from home. I take entire charge of them myself from home, with the help of a man and my 13-year-old boy, except during the extracting season when I have more. To accomplish this I get everything ready at home when there is no work to be done at the out-yards, then the first trip in the spring sees every yard supplied with their supplies for the season. These include fuel for smokers, and even the matches to light them with. The kegs for the honey are taken direct from the factory to each of the yards just before the season opens. Of course I sometimes make a mistake in estimating the amount of store-room required, but it is an easy matter to equalize them when occasion requires. Each yard also has its regular lot of tools and furniture which stays there the year through. This includes the extractor, store-can, uncapping-dish and knife, strainer, and plenty of tin pails for water, etc. There is a supply of nails, screws, racks, wire-cloth, and of course the hammer and screw-driver, two bellows, one automatic smoker, long wisp brooms and a wheelbarrow make up part of the outdoor equipment. In fact, each yard has about everything that is likely to be needed there except the bee-veils, which always go with the man who does the work. The spring locks which are on the houses are all alike, so that one key fits them all. (Coggshall 1898)*

The crew would visit the apiaries on bicycles, since everything they needed was kept there. They would extract the honey, keg it up, and leave it to be picked up later. Sometimes the honey would wait till after the snow fell and they brought it in with horses and a sled.

What really made the Coggshalls unique and established their reputation among the beekeeping community was their descriptions of harvesting honey. The crew was referred to as "lightning operators." They were rough and oblivious to clouds of bees. They employed long whisk (or wisp) brooms which they used to swiftly brush the bees off the honey combs, which were then rushed to a teenager in the tool shed, who extracted the honey from the combs and tossed them back to the lightning operators to be returned to the hives.

I was fortunate to meet one of the descendants of the family, who still kept bees in some of the same locations. One of these still had the shed and he showed hatch marks on the wall noting the number of kegs full of honey that the yard produced each season. The brothers were very prosperous, as evidenced by their large and beautifully built houses, which still stand and are well-kept today. The owners and many of the neighbors know about the Coggshall story, and are usually keen on sharing information. Once, I stopped by Verne Morton's home on the east side of Groton and the owners gladly chatted with me about him. In 2019, my son and two friends bought a farm in West Groton and the following summer we started a small apiary on the property, a mile or so from where the Coggshalls once lived and worked.

Picnic Beekeepers at Coggshalls 1903

For many decades stories by and about the Coggshalls appeared in the beekeeping magazines and even the local newspapers. Here is an excerpt written by one of the hired hands about starting up field work again in the spring:

Part of W. L. Coggshall's beeyard, Genoa, 1908.   *2079*

W. L. Coggshall Beeyard

First Day of Spring Among Out-Apiaries. By Harry S. Howe.

On the morning of March 15, [1898] Mr. W. L. Coggshall and myself were discussing "bee-prospects" by the door of his shop, when I remarkt: "The unusually early spring has started the season for out-door work with the bees from two to three weeks ahead in this locality."

"Yes," said he; "and those south yards should be seen soon."
"But the bottom has dropt out of the roads, almost."

"That is so; but there is no immediate prospects of their being any better, so if you want to take Topsy and the old buggy, go ahead. You can see your own south yards, too, while you are at it."

Getting out a bee-veil I started over the worst possible kind of roads – roads which are a disgrace to the boasted civilization of the State of New York.

Passing through Ithaca, the mile of smooth brick pavement made Topsy forget the 18 miles of mud she had waded through to get there. The Ithaca yard is just west of the city, where there are 107 colonies, still all alive. On the way back from this yard to Ithaca, where I put up for the night, I found a few blossoms of trailing arbutus. (Howe 1898)

Bee-Keepers' Conventions

The *American Bee Journal* was the first publication devoted to beekeeping in the United States; it was formed in 1860 and reported:

We take pleasure in placing on record in our columns, the proceedings of the first American Bee-Keepers' Convention, which met at Cleveland, O., on the 15th of March, 1860. The time is approaching when bee-culture will occupy a higher position than it has yet held in this country, and when it will be interesting to trace back its history to those pioneer movements which conduced to revival and progress. (Newman 1886)

At about this time, the American Civil War took precedence over most activities of this sort. Two conventions were held in 1861, but the publication of the *American Bee Journal* was suspended and Newman relates that "no convention of any importance" occurred again until 1866. The Wisconsin Association formed that year and the "Northwestern" association was formed in 1867 in Des Moines, Iowa. Newman goes on to describe the beginnings of a national organization. Prof. A. J. Cook of Michigan convened a meeting in Indianapolis at the State House of Representatives, held on 21 and 22 December, 1870.

Every seat in the house was occupied ; the States represented being Indiana, Illinois, Michigan, Ohio, Wisconsin, Kentucky, Iowa, New York, Tennessee, Missouri, and Pennsylvania. Delegates were also present from Utah and Canada. On the whole, it is safe to assume that never in the history of America has bee-culture been represented in a convention by so large an assemblage of wide-awake, intelligent, and enterprising bee-keepers. (Newman 1886)

According to the late Cornell Professor Roger A. Morse, the first meeting of New York beekeepers was probably in September 1870, in Utica, NY. He states that detailed information about it is lacking but that Moses Quinby presided. Morse asserts the organization was founded in 1868 and adopted its constitution and bylaws at the aforementioned meeting. According to the *Bee Journal*, the group met again in 1871, and Morse goes on to list subsequent meetings in New York State, including the attendees. Some of them would go on to become large scale producers though at that time, their holdings still appear relatively small. Quinby and his partner Lyman Root (no relation to A. I. Root) claimed 160 hives. Other prominent beekeepers were Solomon Vrooman of Schoharie Co., E. W. Alexander of Oneida Co., and one N. N. Betsinger of Marcellus, who produced 6000 boxes of comb honey from 98 hives. By 1873, Capt. Hetherington was secretary and in 1876 he became president of what was being called the Northeastern Bee-keepers' Association. (Morse 1967)

Worker brood and bees, 1905.   *1088*

Coggshall Bee Gloves

Towards the end of the century, the editor of the *American Bee Journal* asked loyal readers to send word of their experience with the magazine. David Coggshall wrote in reply:

> *FRIEND YORK: – I see you call for a roll of honor, for all who have taken the American Bee Journal for 25 years. I have done better than that; I have taken it ever since it was publisht by A. M. Spangler*

& Co., 25 North 6th St., Philadelphia. Jan. 1, 1861, and I have
nearly every number now. There are a few numbers I have lost
in the mail, and I have neglected to get them. I have been in the
bee-business ever since. I had the first extractor in this county, and
shipt the first extracted honey from this county. I was the second
one to introduce Italian bees. I have over 600 colonies to look after
now, and I produced a big carload of honey in 1898. I taught my
brother, W. L. Coggshall, how to keep bees, and was in partnership
with him several years, under the name of Coggshall Bros. I built
the first movable-frame hives in this section, and have lived to see
bee-keeping advance from almost nothing to what it is now. – D. H.
COGGSHALL. Tompkins Co., N. Y. (Coggshall 1899)

Fig. 28—Coggshall Bee-brush.

Coggshall Bee-brush

Epilogue

The NY Association had a variety of names, finally settling on the Empire State Honey Producers' Assn., in 1935. One of its features has been summer picnics, hosted by prominent beekeepers. In 1920, a picnic was convened at Archie Coggshall's place in Groton. Archie was the son of William Lamar and Sarah Brown Coggshall. Born on January 30, 1887, he died in Ithaca on July 31, 1986 at 99 years old. His brother B. Brown Coggshall was also a beekeeper for many years. In 1940, Brown formed the Finger Lakes Honey Cooperative. He died on May 8, 1976, aged 92.

W. L. Coggshall and sons Brown and Archie in his beeyard, 1897.   525

W. L. Coggshall sons

Archie's son William L. Coggshall was born Dec 1, 1914. William became an Assistant Professor of Apiculture at Cornell University in 1949. He did extension work, was active with the Empire State Honey Producers, and served as president of the Finger Lakes Coop. In 1984 W. L. Coggshall and Prof. Roger A. Morse published the book *Beeswax: Production, Harvesting, Processing and Products*. He died in 1986 and was buried with his extended family in Groton, NY.

Millard Vernet Coggshall was born on March 27, 1915 in Groton, NY, the son of B. B. Coggshall and Jennie Steinberg Coggshall. Millard graduated from Groton High School in 1933 and received a Bachelor of Science in Agriculture from Cornell in 1937. Millard and his wife moved to Minneola, Florida in the early 1940's where he established Coggshall's Honey Business. He was president of the Florida State Beekeepers Association and received the Florida State Beekeepers Pioneer Award in 2002. Millard Coggshall passed away August 29, 2006.

Works Cited

Anon. (1884). *American Bee Journal*. Vol. 20. No. 31.

Coggshall, D. H. (1899). *American Bee Journal*. Vol. 39. No. 32. Coggshall, W. L. (1898). *American Bee Journal*. Vol. 38, No. 36. Howe, H. (1898). *American Bee Journal*. Vol. 38, No. 22.

King, A. J. (1879). *The New Bee-keepers' Text-book*. O. Judd Company.

Morse, R. A. (1967). *A Short History of the Empire State Honey Producers' Association*. Office of Apiculture, Department of Entomology, Cornell University.

Newman, T. G. (1886). *Brief History of the North American Bee-Keepers Society*. Office of the *American Bee Journal*. Chicago, Ill.

Chapter 13

How the Honey Bee Came to Mexico

Cultiva las Colmenas, según te lo previene Salomón en sus Proverbios; admirarás en el manejo de esos pequeños animales la disciplina de los Monasterios, y la política de los Reyes.

Cultivate the Beehives, as Solomon warns you in his Proverbs; you will admire in the handling of these small animals the discipline of the Monasteries, and the politics of the Kings. (de la P laza 1797)

The honey bee is not native to the Americas, it was brought here from Europe. Looking into the story of how they were brought to the United States, I found many surprising things. Detailed records of the many Atlantic crossings are not easily obtained, either because they hadn't been written or were lost over time. Perhaps few people thought about leaving something to tell the story centuries later. Further difficulty arises from detailed descriptions of doubtful accuracy.

Many observers saw honey bees living in the woods of New England and concluded they must be native, whereas these bees were descendants of early imports. Finally, a lot of plain rubbish was written about the New World being a land of milk and honey, whereas in truth it was more like mud and hunger. When the British threatened to take New Amsterdam from the Dutch – what is now New York – the residents thought of that as an improvement. Things could hardly get any worse than they already were.

So, it came as no surprise to find a similar lack of information, erroneous descriptions, and speculative narratives regarding the honey bee in New Spain, which later became what is now Mexico. No doubt confusing the issue was the fact that there were already bees that store honey there. These are technically not honey bees (*Apis mellifera*) but the so-called stingless bees which comprise the genera *Melipona*, *Trigona*, and so forth. These bees form colonies, have one or more queens and a myriad of workers. They also produce wax from glands in their bodies but they mix it with plant resins, fibers, and dirt, rendering it black and gummy. The honey is stored in little pots made from this material (*cerumen*). Stingless bees produce much smaller amounts of honey than the European honey bees.

The legends begin with Columbus being handed a chunk of wax in the West Indies. The story was later related by Alexander von Humboldt in 1856:

> Columbus expressly says, that in his time the natives of Cuba did not gather wax. The great cake of this substance which he found in the island on his first voyage, and which was presented to King Ferdinand, to the celebrated audience at Barcelona, was found afterwards, to have been brought by Mexican pirogues from Yucatán. It is curious to observe that the wax of the Melipones was the first Mexican production that fell into the hands of the Spaniards, in the month of November, 1492. (Humbolt 1856)

The Spanish invaded what is now Mexico, as well as most of the countries to the south. They found many established civilizations for which they had little or no respect and attempted to impose the Spanish Language and Christianity on the citizens while exploiting the people and the resources. One of these resources was the vast honey and wax culture of the Mayan people living in the Yucatán. Traditional beekeeping is still practiced there and was studied extensively by Harriet de Jong. In her 1999 book *The Land of Corn and Honey* she describes the culture of *Melipona beechii*, which the Mayans call *Xunan kab*.

> In the homestead, people keep Xunan kab in horizontal hollow logs stacked up in a special bee-house. The logs are never placed upright: the beekeepers argue that the Xunan kab of the forest live in horizontal branches of trees and therefore do not like to live in a vertical hive. Xunan kab produces honey of incomparable quality. The Maya explain this by saying that Xunan kab goes to Xmaben to collect honey. Xmaben is a celestial field of flowers. As this is the only species which has access to Xmaben, its honey is said to be divine in both origin and quality. Some people also acknowledge that Xunan kab collects nectar and pollen from earthly flowers, but only from a few of the finest species. (de Jo ng 1999)

The Honey Bee In New Spain and Mexico

Notwithstanding the presence of this widespread centuries old tradition of beekeeping with stingless bees, the story has become widespread that the Spanish, upon not finding European honey bees in New Spain, proceeded to import them as the Dutch, English and Germans had apparently done in North America. Despite the flimsiest of evidence, Donald Brand created an elaborate

Traditional hive, Spain

scenario complete with many pages of sources and citations. In his oft-quoted *The Honey Bee In New Spain and Mexico* he said:

There was an introduction of the European bee into New Spain in the 1520s or 1530s. The early colonial Spaniards and creoles in New Spain were not bee-keepers, the highland Indians were poor beekeepers, and soon most of the European bees were quite wild. The European bee went with the religious orders (especially the Franciscans and Jesuits) into northern Mexico, were introduced to the Indians, and eventually went wild in the Sierra Madre Occidental and Sierra Madre Oriental, where today [1970] there is the greatest concentration of wild small black European bees. (Brand 1988)

According to this story, the Spanish carried the honey bee on to Arizona and California where pockets of wild black bees still exist. In a 1925 University of California publication, "Survey of Beekeeping in California", George Vansell adds another page to the tale when he states the "Spanish Bee" of California is "thought to be this same old German type which was brought into the state in early days by the Spanish Fathers."

Going back to Donald Brand, he suggested "The Spaniards, with their traditional craving for sweetmeats, probably found the supply of honey that reached them in the Mexican highlands insufficient to supply their needs." He also bolsters his tale by insisting that this and the need for beeswax by the church, gave impetus to the importation of honey bees from Spain into Mexico, and its environs. Both of these suppositions can be easily disputed.

As we know, the indigenous people of Mexico had extensive civilizations and they cultivated a variety of plants, including corn (maize) and agave (maguey). According to Alexander Humboldt:

Before the arrival of the Europeans, the Mexicans and Peruvians pressed out the juice of the maize-stalk to make sugar from it. They not only concentrated this juice by evaporation; they knew also to prepare the rough sugar by cooling the thickened syrup. Cortez, describing to the Emperor Charles V all the commodities sold in the great market of Tlatelolco, on his entry into Tenochtitlan, expressly names the Mexican sugar. "There is sold," says he, "honey of bees and wax, honey from the stalks of maize, which are as sweet as sugar-cane, and honey from a shrub called by the people maguey. The natives make sugar of these plants, and this sugar they also sell." (von Humboldt 1856)

Not only was there an abundance of sweeteners to satisfy everyone's craving for sugar, but the people were fully capable of producing alcoholic beverages from the surplus of sugary products. Humboldt: "Under the monastic government of the Incas it was not permitted in Peru to manufacture intoxicating liquors [but] drunkenness was very common among the Indians of the times of the Aztec dynasty."

Beeswax in the New World

Fig. 1. End view of log hives of colecab *(M. beecheii)* stacked on an A-frame rack under a thatched roof. An empty hive is in the foreground near the shelter. Rocks on which hives are balanced when honey is removed from them can be seen in front of each bank of hives; the one on the right had been used as a watering trough. The plants with the spear-like leaves in the near background are henequen (sisal).

Beekeeping with the Stingless Bee by the Yucatecan Maya

Regarding the need for beeswax in the New World, we come to a very curious irony. Donald Brand expounded on this at great length, but to be brief:

> *Possibly the greatest reason for introducing the European bee was the fact that the Roman Catholic Church placed a mystic and spiritual value on the candle made of white beeswax, in addition to a natural dislike of the messy and dirty candle made from the native Cera de Campeche [stingless bees' wax]. (Brand 1988)*

It is true that the Catholics had a voracious appetite for pure beeswax. It was revered because of the supposed virginity of honey bees (no one had figured out the role of the drone bees, with which the queen mates with many over several days, causing the drones to drop dead); beeswax represented the body of Christ, born of a virgin mother (Brand 1988). The logical assumption is the demand for beeswax would lead to beekeeping in America, but according to a 1988 article by Robert Kent:

> The first introduction of the honeybee to the Western Hemisphere occurred in North America before the end of the sixteenth century. Honeybees were numerous in the countryside of the English colonies of Virginia and Massachusetts by 1650. The Spanish, on the other hand, successfully kept bees from being introduced into their colonies for many years. This apparently was done to guarantee the New World ecclesiastical candle wax market for beeswax produced in Spain. (Kent 1988)

Robert Kent cites La Plaza, "Memoria Sobre La Cria De Abejas Y Cultivado La Cera (Memoir on Bee Culture and the Cultivation of Wax)." This short treatise was penned by "El Bachiller Eugenio de la Plaza (The Bachelor Eugene de la Plaza)" in Cuba and published in 1797. The copy I have includes a typewritten summary and explanation in English, which is helpful since I am not fluent in 18th century Spanish. Taken from that work: "In the year 1795 the Cuban Government offered a prize of 300 pesos for the best Essay on bee-keeping and the production of wax." The winner was La Plaza, a "writer on scientific matters," who died in 1804 in Havana. Colonel H. J. O. Walker has summarized the key points:

> At the time of the Spanish conquests this insect had not been established in America, and its later introduction was from Europe. In la Nueva España and other internal Provinces adjacent there are, says Plaza, no bees, while in Florida they are very plentiful, so that it is natural to conclude that they were introduced there by the English. The introduction of the honey-bee into Cuba was modern. In 1763, when Spain ceded Florida to the English, many of the inhabitants of the city of St. Augustine transferred themselves to Cuba, taking with them their stocks of bees. Some of these escaped and became established in the mountains, where they throve and multiplied greatly. (Walker 1910)

Typical Apiary of Modern Hives in Mexico

Contrary to the conventional wisdom, the people of New Spain were not producing their own beeswax with Spanish bees at all, but were importing huge quantities of it. According to Kent, Spain prohibited beeswax production, probably to keep New Spain from becoming self-sufficient as well as maintain a balance of trade where they needed valuable items from Old Spain for which they traded items they produced. When the honey bee was finally imported into Mexico, the bees came not from Spain but from the United States. The Spanish swapped Florida for Cuba, and the citizens of St. Augustine took bees to Cuba where they flourished. Seeing how well they did, bees were then shipped to Mexico. But even then, the Yucatán resisted due to the well established industry of stingless beekeeping. According to Charles F. Calkins, in his 1974 dissertation on "Beekeeping in Yucatán":

> The initial appearance of Apis mellifera in Mexico, as the best
> evidence suggests, was shortly after its introduction to Cuba. There
> is no evidence to indicate the presence of Apis mellifera in the
> Yucatán area between 1764 and 1855. The first introduction of Apis
> mellifera to Yucatán took place only a relatively short time ago. The
> exact date cannot be determined with any certainty. In general
> terms, however, the introduction was at about the turn of the 20th

century. It is interesting to note that this bee is commonly referred to as "la abeja americana," the American bee. This nomenclature, of course, suggests that the bee was introduced from the United States, not elsewhere in Mexico. In the context of its usage in Yucatán, the term "americana" refers to the United States. (Calkins 1974)

Calkins goes on to say that the "race" (subspecies) of that importation is not known, although he suggests the likelihood of *Apis m. mellifera* (the dark or German bee) or perhaps *A. m. caucasica* (also dark). The first documented import was this latter although over time, *A. m. ligustica* (the Italian bee) became predominant. The introduction of the European honey bee revolutionized beekeeping in the Yucatán, of that there can be no doubt. In a few decades the density of honey bee colonies was as high or higher than anywhere in the world. The production of wax and honey from *Apis* colonies far exceeds that of stingless bees, and many traditional beekeepers adopted the "American bee." Sadly, it has had a negative impact on traditional crafts, values, and culture. Add to that the deforestation and conversion to agriculture, the indigenous lifestyle is under threat of extinction. Especially sad because the Mayan culture persisted during the reign of terror brought by the Spanish and later the exploitation by Mexican culture which continues today, to finally succumb to large scale agriculture, of which it can take no part.

Genealogical Research

The question of the genetic background of the bees of Mexico has now been introduced. Like most genealogical research, there has been a great sea change from the poring over family trees and ship logs, to the use of DNA. I assume most readers have heard of companies like 23andMe, Ancestry.com, and so on, that claim to be able to tell you where your roots are and what nationalities and ethnicities lurk in your background. Personally, I discovered my background was far less German, despite my surname *Borst*, and predominantly from the British Isles. I should have known since my mother's family is made up of Youngs, Campbells, Lorings and Pratts. Sometimes you only find what you are looking for. The chief problem with DNA research is that an organism's genome is vast and filled with information, most of it not relevant to the inquiry. We have to go in with a definite plan of "what we are looking for." Otherwise, it would be like going to the library to find a book on a certain topic by just walking up and down the aisles (this could take forever). Has anyone really looked at this exact question: were the bees of Mexico from Germany or Spain? I asked my friend Dr. Ernesto

Guzman, from the University of Guelph. He was born in and worked with bees for many years in Mexico. He said: "The short answer to your question is no, I do not have knowledge that *A. m. iberiensis* markers have been studied in Mexico or Southern USA. My understanding is that the first colonies introduced in Mexico were imported from Cuba, likely around 1760."

As early as 1988, scientists were beginning to use DNA analysis to determine the genetic background of honey bees. This was spurred by the disastrous importation of bees from Africa into Brazil. I say disastrous, because this was an entirely unexpected result of a breeding experiment which led to the honey bee becoming an invasive pest in most of tropical America. In some circles it is viewed as a benefit, as the Africanized bees are apparently healthy and productive in the tropical climate, which European bees were not as much (hence the idea of crossing them with tropical African bees). Unfortunately, the hybrids tend to be vicious, killing livestock and sometimes people.

One of the first mentions of this tactic is in "Polymorphisms in mitochondrial DNA of European and Africanized honeybees (Apis mellifera)" by D. R. Smith and W. M. Brown. I realize that is a mouthful, but to put it simply, most animals have a mitochondrial DNA sequence which is much simpler and smaller than the DNA that makes up the whole genome. Parts of this sequence are unique to each species and can even identify particular subspecies. The downside is that "mtDNA" is only inherited from the mother, so you miss some important information. But it was viewed early on as a tool for differentiating between African and European lineages. Essentially, honey bees break into several lineages, chiefly Northern European, Mediterranean and African, as represented by: 1) English, French, and German bees; 2) the Italian and Carniolan lines; and 3) African bees which range from Egypt to Algeria, and the rest of Africa.

Unfortunately, these assays cannot tell us everything. For example, if we are looking for bees with an African genetic marker, it may not tell us anything about what part of Africa the bees came from. In the early days of genotyping, Kandemir *et al*, discovered that in fact some bees in the United States are descended from *Apis m. lamarckii*, which had been imported briefly in the 1800s – from Egypt. So, a test for Africanization would have to be able to discriminate Egyptian mtDNA, or lead to a false positive.

Deep Dive into the DNA

A 2017 article in the journal *Apidologie*, titled "Mitochondrial DNA variation of *Apis mellifera iberiensis*" (Chávez-Galarza et al), refers to the native bees of Spain and Portugal. The authors state that "Mitochondrial DNA (mtDNA) has indisputably been the most popular and longest used marker in Iberia." Ruttner, in 1988, considered the Iberian bee to be firmly in the lineage with the rest of Northern Europe (called the M lineage), based on physical characters. However, many of these same characters are found in North African bees. Ultimately, the authors concluded that Iberian bees represent a hybrid of European and African lineages: "Colonies of A ancestry were predominant (67.5%), as compared with those of M (32.2%) and C ancestry (0.3%)." Finally, they note that "Iberia is an important source of honey bee mtDNA diversity especially of African ancestry."

Most recently, Erin Calfee and her colleagues reported their findings in *PLOS Genetics*:

> *Colonists as early as the 1600s imported European honey bee subspecies … setting off the first honey bee invasion of the Americas. Historical sources indicate that the A ancestry is from A. m. scutellata, while both M and C ancestries are mixtures of multiple subspecies imported from Europe, e.g. A. m. ligustica (C), A. m. carnica (C), A. m. mellifera (M), and A. m. iberiensis (M). (Calfee 2020)*

The authors anticipated that these lineages would remain linked in the hybrid so-called "Africanized" bees. Unexpectedly, while the African and Eastern European lines were consistent throughout the range from North and South America, the Western European DNA (M lineage) broke into two populations. Bees in the temperate regions are genetically linked to the Northern European strain (*Apis mellifera mellifera*) while the equatorial regions show linkage to the bees from Spain (*Apis mellifera iberiensis*).

Early picture of extracting honey in Mexico

While this could be the result of natural selection on honey bee types according to climate, they assert that it is a reflection of the background of the original importations of bees. To quote Calfee *et al*: "M ancestry at lower latitudes in South America is more similar to *Apis mellifera iberiensis* (Spain) than M ancestry elsewhere in the Americas." In other words, the initial importations left a mark on their descendants. While diluted by having been crossed with Italian stock and later swamped by African DNA, the mark remains. Unfortunately, this group did not include samples from Mexico in their analysis. This, as they say, could be pursued in future research – which might shed more light on the question of from where the honey bee first came to Mexico.

Works Cited

Brand, D. (1988). The honey bee in New Spain and Mexico. *Journal of Cultural Geography*, 9(1), 71-82.

Calfee, E., et al. (2020). Selection and hybridization shaped the rapid spread of African honey bee ancestry in the Americas. *PLOS genetics*, 16(10).

Calkins, C. (1974). *Beekeeping in Yucatan* (Doctoral Dissertation, University of Nebraska).

Chávez-Galarza, J., et al. (2017). Mitochondrial DNA variation of Apis mellifera iberiensis: further insights from a large-scale study using sequence data of the tRNA leu-cox2 intergenic region. *Apidologie*, 48(4), 533-544.

De Jong, H. (1999). *The land of corn and honey: The keeping of stingless bees (Meliponiculture) in the ethno-ecological environment of Yucatan (Mexico) and El Salvador.* Universiteit Utrecht.

De La Plaza, E. (1797). *Memoria sobre la cria de abejas y cultivo de la cera premiada por la Junta de Gobierno del Real Consulado de la ciudad de la Havana.* Havana : Por D.E.J. Boloña

Kent, R. (1988). The introduction and diffusion of the African honeybee in South America. *Yearbook of the Association of Pacific Coast Geographers*, 50(1), 21-43.

Ruttner, F. (1988). *Biogeography and Taxonomy of Honeybees*. Springer Berlin Heidelberg . Smith, D., & Brown, W. (1988). Polymorphisms in mitochondrial DNA of European and Africanized honeybees (Apis mellifera). *Experientia, 44*(3), 257-260.

Vansell, G. H. (1925). *A Survey of Beekeeping in California* (Vol. 297). University of California, College of Agriculture, Agricultural Experiment Station.

Von Humboldt, A., & Thrasher, J. S. (1856). *The island of Cuba*. Derby & Jackson. Walker, H. J. O. (1910). Summary (In English) of De La Plaza, above.

Kandemir, I., Meixner, M. D., & Sheppard, W. S. (2003). Morphometric, allozymic, and mtDNA variation in honeybee (Apis mellifera cypria, Pollman 1879) populations in northern Cyprus. In 38th Apimondia International Apicultural Congress.

Chapter 14

Stories of Beekeeping in San Diego

Our family moved to San Diego County in late 1959. To me, it was a great shock – having been born in Boston and living the first nine years of my life on tree-lined streets in the suburbs. Southern California struck me as a great wasteland, punctuated by groves of orange trees, eucalyptus, and the ever present prickly pear cactus. After 9 years on the West Coast, I definitely had the bi-coastal disorder. I traveled back and forth from the beaches of the East and West, for many years, until finally settling in my forties among the Finger Lakes of New York.

I caught the beekeeping bug in 1974 through reading. My first encounters were Maeterlinck's *The Life of the Bee*; a counterculture guidebook entitled *One Acre and Security*; and a very funny short story by John Barth, about how a baby was named Ambrose on account of a swarm of bees. The second one was strictly practical, although it presented its subjects in an overly optimistic light. Still under the sway of the bi-coastal affliction, my new wife and I traveled to Ithaca NY, home base of the eminent Cornell University entomology professor Roger Morse. His advice was succinct: "Get a job," he said, and so I did – working for a beekeeper who managed more than 2000 hives without permanent help.

The experience of going from zero to 2000 in a week was eye-opening, to say the least. By late summer I had grown weary of 70 hour weeks and endless stinging episodes where I ended up covered with welts. Baptized by fire, is the way I recall it. The young couple, and a young one well on the way, retreated to San Diego County where I began a string of jobs working for various beekeepers. Ultimately, I got a job at a small factory owned by Henry Knorr, where we manufactured beeswax foundation for hives, and 31 different colors of beeswax candles. We also had a small store where we sold beekeeping equipment, and bought wax by the ton. Approximately two tons were processed every day, going about equally into candles and wax sheets for the beekeepers.

Pretty soon I knew the names of all the prominent beekeepers on the West Coast and beyond, and began to build up my own apiary. I learned about the phenomenal crops of black button sage (*Salvia mellifera*), a water white honey that poured in after the winter rains. And of the rich buttery eucalyptus honey that was gathered by the ton, even when it didn't rain. I was hypnotized by the fragrance of orange blossom nectar, which always seemed to evaporate between extracting the honey and getting it into jars. Orange blossom nectar is mostly water, so when the bees concentrate it, there isn't much left and is usually overpowered by the eucalyptus.

Ferdinand Knorr fills an order for sixty pair of pressed-out honeycomb candles in an assortment of fascinating colors and lengths for ex-vice President Barkley.

1954 Ferdinand Knorr

One of the first books on practical beekeeping that I acquired was Root's *ABC and XYZ*. Naturally, I obtained the latest edition, but I also got a reprint of the 1890 edition, as well as a reprint of the classic *Hive and Honey-Bee* by Langstroth, from 1853. So, early on I began to see modern beekeeping as a continuum which began in earnest in the 1800s and continues to the present day.

Early Days

The story of John Harbison has been told many times and I will just touch upon it briefly. Joseph Bray summarized it in his preface to a reprint of William Harbison's 1860 book:

> In 1869 John moved from Sacramento to the virgin bee ranges of San Diego County, and soon made newspaper headlines with the trainloads of California honey he shipped to Chicago and New York. One year he harvested 60,000 pounds of comb honey from 300 colonies of bees. Within ten years of moving to San Diego John S. Harbison became the largest honey producer in the world, with nearly 4000 hives in apiaries throughout San Diego's sparsely settled backcountry. For a time in the 1870s San Diego reigned supreme as the top honey-producing county in the United States, largely due to Harbison's enterprise. (Bray 2012)

The jaw-dropping stories naturally caught the attention of other would-be fortune seekers whose luck may or may not have panned out in the gold fields of northern California. By this time, Harbison had moved on to more reliable sources of income such as the grocery business and real estate. It soon became clear to all that the phenomenal crops of water white sage honey were wholly dependent on ample rainfall, which often failed to materialize – sometimes for years in succession. When large scale irrigation projects began, often using vast quantities of redwood timber for flues and aqueducts, fruit growing began in earnest. It is one of the ironies of life that fruit growers and beekeepers entered into a pitched battle, with the growers claiming that honey bees ruined their fruit. Whole apiaries were set ablaze and the honey producers were often driven up narrow canyons too steep to cultivate but blooming in profusion with a wide array of herbs and shrubs.

San Diego has a typical Mediterranean climate, so it's suitable for growing grapes, oranges, pomegranates and the like. I became interested in botany as a teenager and was surprised to learn that while we had plants belonging to the same families as Mediterranean ones, the actual species were very different. While so many culinary herbs grow natively in France, Italy and Greece, very few equivalents can be found native in San Diego. True, there is a variety of sages (*Salvia*), but they are all too bitter to be used to season a meal. Also, many types of sumac (*Rhus*) can be found, but they only resemble the species from other parts of the world. But despite the fact that the honey bee is not native to

California, many species of flowering plants co-evolved with the native wild bees that still live among the hills and canyons of San Diego County.

My former boss, Henry Knorr, told stories of the primitive beekeeping in his early days. His father Ferdinand was born in 1885 and emigrated from Poland to the US in 1904, finally settling in the rural outskirts of San Diego. Henry said of life in the 1920s, "There was no electricity or indoor plumbing. It was like camping and I never cared for camping." Henry learned to dislike beekeeping because he and his father used to extract honey in the bee yard, inside a tent. The honey extractor would be set right upon the ground. It had to be frequently emptied into 5-gallon cans, which they set in a hole in the ground, below the spigot of the honey extractor. Henry learned the machinist's trade like his father, and moved to the city of San Diego, hoping to leave the bees behind. But his father Ferdinand had built a prosperous beeswax business and eventually, Henry took it over from him. Both Henry and his father lived into their late 90s.

Messrs. Morgan and Morse

1870 Rufus Morgan

Rufus Morgan was eking out a living as a beekeeper in North Carolina when in 1878, he heard of the California bonanza. According to an article published in 1994, in *The Journal of San Diego History*, Morgan left his pregnant wife and three year old daughter, and traveled by train to Chicago, on to San Francisco, thence boarding a ship to San Diego. He carried a letter of introduction which informed the receiver that he was "an Apiarist, and standard authority throughout the United States, on the subject of Bees." He entered into an agreement with a wealthy businessman named Ephraim W. Morse, whose fortunes followed a far different path than Mr. Morgan's.

With Morse's assistance, an apiary was established in the "Oak Glen" which was in the vicinity of present-day Rancho Bernardo. The sad story is rather fully described in the letters which make up the bulk of the historical piece, in which Morgan's plans were abruptly terminated by a fatal meal of wild mushrooms, leaving his wife and family penniless. In fact, the illustrious Mr. Morse sent her a letter claiming to be owed a lot of money by the deceased. But the letters 1890

Beekeeping ranch

themselves present a vivid story of what beekeeping was like in those days. For example:

Jan. 24, 1879. The honey prospect is good and I will be able to get all the bees I wish on shares but I will not be able to make as much honey as I expected to, for you see I will have to divide profits – but as the business is even better than I anticipated before I came, that can be put up with for one season. I have had lots of offers – a dozen men are eager to secure my services, but I will be in no hurry to tie to anyone. Have also offered me all the money I will need in case I wish to purchase – of course to divide profits. The honey business is a big thing here, bigger than I had any idea of – I know in five years from now I will be making $5000 per year! – Fortunes are made at it here & all are doing well at it. In a word, I am more than pleased so far, and feel sure we have a certainty of a happy home here and all the comforts that we'll need. Its a much better honey country than I expected – every shrub and bush and sprig of grass, if you ask the name of it, they will say – "thats one of our best honey plants"

June 22, 1879. I have the bad news to tell you that the prospect is for no honey – this years disastrous failure will break hundreds up, including many merchants: Its the 2d bad year they have ever had, and people feel it keenly – over 40 families left San Diego last month & nice houses can be had for $5 & $7 per month. Of course

I lose nothing but my time, but that is something, but will be repaid by the increased price of honey. If we make a good crop next year – the increase in price will equal what we lose this.

Nov. 17, 1879. A perfect gale from the N.E. is blowing and I would not be surprised to see more rain, as the weather is so unsettled – every thing points to a very wet season – though not a very large honey yield for the county on account of fire, which has left a space of 75 miles as bare as a floor.

April 4, 1880. The rainy season is, or ought to be over yet we had two days rain since my last – so that every one agrees this will be a memorable year. (Morgan 1994)

This was his last letter. Mrs. Morgan received a note informing of her husband's sad demise and a bill for $125. Morse offered to return his remaining possessions: "a gold watch chain, two gold studs, a few books, a trunk, some clothing, and a quantity of pressed ferns."

The Eucalypts

As a boy, growing up in the 1960s, I spent much time wandering about the hills and canyons of Rancho Santa Fe. We learned in grade school that the Santa Fe railroad introduced the eucalyptus tree to California, and by the time I heard this story, there were vast forests of the pungent, sappy trees. Even then, some of them had attained an awesome hugeness. We sat under their shade and listened to them creak and moan in the breeze. As a beekeeper in the 1970s, I learned what a boon the trees were for making honey. I met one beekeeper who owned a single hive in the Point Loma neighborhood of San Diego. He told me that he could obtain with virtually no effort, more than 300 pounds a year from his one hive, principally from eucalyptus, but also from the variety of wild and cultivated flowers in the urban sections of the county.

Like many beekeepers, I tried to catch the hit or miss flows from the wild sages, which produced honey of astonishing clarity that is reputed never to granulate but stay liquid for years. The eucalypts were far more reliable, catching moisture from the humid ocean breezes and the water poured into lawns and gardens. I remember one vast wooded canyon that was a sanctuary for the annual migration of monarchs, making it a magical spot of graceful deep green boughs and the fluttering orange wings of the butterfly swarms.

Eriogonum fasciculatum

Salvia mellifera

As early as the 1870s, eucalyptus was being promoted as a honey plant. This amusing piece appeared in a local newspaper:

> "The Eucalyptus as a Honey-producing Tree," is the title of an article in the Pacific Rural Press, "written by U. K. Lyptus," and closing as follows: "In addition to the honey yield of the blossoms, the eucalyptus will serve the apiarian as a wind-break; it will furnish him fuel and lumber for hives, stands, boxes and buildings. The dried leaves furnish excellent fumes to the smoker. We have satisfied ourselves that the eucalyptus is indispensable around an apiary and we hope others will investigate the matter to their own satisfaction." If there is anything the eucalyptus is not good for, won't somebody kindly name it?

As I mentioned, I worked in a bee supply factory in the 1970s. This establishment was located at the western edge of Rancho Santa Fe, a town where thousands of acres of eucalyptus grow. One time, an Australian visitor dropped in to the store. He was dumfounded! He reiterated the idea that eucalyptus trees could be used for almost everything, but with the additional information that for each use, there was a particular species or variety best suited. "Eucs" for flooring, lumber, furniture, and so on. But, he exclaimed, the one we had – the blue gum – was good for nothing! I knew he was right, we learned in school that the railroad company picked the fastest growing tree in order to have logs on which to lay their train tracks. Blue gum grows fast, for certain. However, when it was green the railroad spikes would not stay in the wood; they backed out as the wood dried. When the wood was properly dried, it became far too hard to drive the spikes into it.

I eventually had almost 500 hives of bees in and around this oasis. The blue gum is a fine source of nectar and pollen, though it begins blooming in late December so it was often regarded as a forage plant to build bees up on. Savvy San Diego honey producers would move their apiaries, usually numbering 120 or so per truck load, to the black button sage, in spring, and up to the mountains to catch the wild buckwheat flow (*Eriogonum fasciculatum*). In the years when rainfall was poor, they would truck the hives down into the Imperial Valley desert, to try to make a crop from irrigated alfalfa.

Bees on Wheels

1975 Pete and his Bees

One of the first beekeepers I worked with was Jim Austin, who was based in the desert. In early spring he brought his hives over from the desert to the coast, to build up and make splits. There was a profusion of bloom: wild California "lilac" (*Ceanothus*), mustard (*Brassica*), and filaree (*Erodium*). I remember loading and unloading the hives by hand, all weather-beaten and gray like pieces of sun-bleached desert wood. He set the live hives down in long rows, and put vacant hives in rows directly behind them. Then we set the top box of bees from the live hive onto the vacant hive, one after another until the numbers were built back up. He would move them back to the desert in June to make honey. "I might get a decent crop before the aerial spraying kills 'em," he told me. Jim's father had been a beekeeper before him. The whole family would travel to the coast and camp on the beach while they tended to the hives, resurrecting them from the grueling desert summers of temperatures over a hundred for weeks and weeks. The hives in the desert have to be placed under shade structures (*ramadas*) or the wax melts and runs out the front.

Migratory beekeeping has a long history in California, stretching back to the 1870s. According to Clifford M. Zierer, writing in 1932, the practice began when beekeepers moved their hives from the sage and buckwheat ranges to cultivated lima bean fields on the coast, in order to avoid having to feed the bees in times of scarcity. He stated:

> *Hundreds of colonies were hauled from 20 to 40 miles on specially constructed wagons drawn by four or six horses. The moves were made at night in order to avoid the midday heat and consequent softening of the combs and smothering of the bees. It was the general opinion at that time that migrations were worth while only in case the sage crop failed. (Zierer 1932)*

By the time I got into beekeeping, commercial beekeepers were well mobilized. In the 1970s, many of these had electric boom loaders which could pick up one or two hives at a time. These trucks typically held about 120 hives or what was considered a single apiary. If a location could not support that many hives, it was not worth using it. But many still loaded by hand, hiring help as needed. Naturally this was back-breaking labor, hefting 100 or more pounds at a time onto a chest-high flat bed, and then stacking them up. Jim Austin told me a closely guarded secret. He said he had a very burly guy helping him load hives, day after day. Finally, the man said to Jim, "Why am I so tired, and you seem to be doing fine?" He replied that he would tip the hive towards the other man so the center of gravity would shift and his share of the lifting was less as a result.

The Bees of California

Growing up in San Diego, we all learned of the Spanish Missionaries who built churches from Mexico clear up to Northern California, along "El Camino Real" or The Royal Road. We visited the mission at San Diego, learned of Father Junipero Serra and how he brought Christianity to the natives. Somehow, over the years, the story arose that these pioneers brought honey bees with them. In a University of California publication, George Vansell said bees were thought to have been "brought into the state in early days by the Spanish Fathers." However, in 1878, a writer in the *American Bee Journal* clarifies:

> *The first settlers, in 1760, were the catholic missionaries who did most remarkable work introducing the products and appliances of civilization. But among the many things brought by them, there is no record of the honey-bee, nor did any of the Spanish native residents of California know anything of bees or honey until after the discovery of gold and the influx of "Americans," as people coming from our Eastern States were called. (American Bee Journal 1879)*

San Diego has changed greatly since it was first seen by Cabrillo in 1542. As boys we hiked along the vast road grading that was being done to make way for the Interstate highways in the 1960s; now there are innumerable highways running up and down the beautiful valleys where beekeepers once set their hives to capture the essence of the wild sage and buckwheat. The bees have changed, too. These days if someone sees a swarm of bees hanging on a tree branch, it is safely assumed they are invasive African bees.

Works Cited

American Bee Journal. (1879) Vol. XIV. No. 12.

Angier, B. (1972). One Acre and Security: How to Live Off the Earth Without Ruining it. Stackpole Books.

Barth, J. (1963). Ambrose His Mark. Lost in the Funhouse

Bray. J. (2012) in: reprint of Harbison's California Adjustable Comb Hive: Patented by JS Harbison, January 1, 1859.

Maeterlinck, M. (1901). The life of the bee. Dodd, Mead.

Morgan, R. (1994) Letters Written from San Diego County, 1879–1880. Journal of San Diego History, 40, 142-77.

Morning Press. (1878) Santa Barbara. Volume VI, Number 304, 21 May

Zierer, C. M. (1932). Migratory beekeepers of southern California. Geographical Review, 22(2), 260-269.

Chapter 15

Early Migratory Beekeepers - Better Pasture

Moving Bees in New Zealand

1907 Moving Bees in New Zealand Gleanings

The idea of moving bees to better pastures is as old as beekeeping itself. Supposedly, hives were floated up and down the Nile on barges by the ancient Egyptians. In 1890, the beekeeping journal *Gleanings in Bee Culture* reported that beekeepers carried hives on their backs, up into the Austrian Alps to take advantage of summer pasture in the mountains.

> *The Carinthian bee-keepers, to secure the highly prized honey of the Alps, carry, in mid-summer, with much labor, their stocks of bees to the highest Alpine meadows. The cases are carried upon the back, in what are called "knaxen," and more than three of them are sometimes piled on; and when we consider the weight of the cases, and the very difficult road, the burden seems possible for only the strongest men. One occasionally sees the entire bee-keeping guild of the Carinthian high-mountain villages, laboriously threading their way along the narrow paths, climbing upward, as shown in the cut.*

> *What bee-keeper's heart is not stirred with enthusiasm in looking
> at the picture, and seeing the Carinthians climbing up the cloudy
> mountains? (It says that only the strongest man can carry three
> colonies; but it seems as if the lady had the largest load, does it
> not?) (Gleanings 1890)*

Prior to the invention of motorized vehicles, moving large numbers of hives was
done with horse drawn wagons. Jacob Biggle of Pennsylvania, in his 1909 book,
gives a description of what beekeeping was like in those days:

> *There are some beekeepers who each year move their apiary from
> place to place, following the bloom; but as this method is precarious
> at best, I should not advise its adoption except in rare instances,
> and even then only in the hands of an expert. In moving bees to and
> from the out yards, a great deal of caution should be exercised; for
> should the hive be rudely jarred and the bees escape while in transit,
> disastrous consequences will surely follow, and possibly the horses
> may be stung to death. (Biggle 1909)*

California, Well on the Way

Mr. Biggle may have advised against it but according to Clifford Zierer, by this
time the beekeepers of Southern California were well on their way to migratory
beekeeping:

> *Before 1895 a few beemen moved their apiaries from sage and wild
> buckwheat ranges to lima-bean fields in dry years in order to avoid
> the expense of feeding. Soon after 1895 the practice became rather
> common in Ventura and Los Angeles counties. Hundreds of colonies
> were hauled from 20 to 40 miles on specially constructed wagons
> drawn by four or six horses. The moves were made at night in order
> to avoid the midday heat and consequent softening of the combs
> and smothering of the bees. Some apiaries were moved a few
> miles from sage to wild buckwheat at higher elevations in the coast
> ranges. It was the general opinion at that time that migrations were
> worthwhile only in case the sage crop failed. (Zierer 1932).*

Lee Watkins, who was a beekeeper in the early twentieth century, wrote
extensively about the history of beekeeping in California. He reported:

It appears that Mr. William Buck, California's first large importer of honeybees, was also its first migratory beekeeper. During the early summer of 1858 he loaded 48 hives on a wagon in his apiary just south of San Jose and took them to the San Francisco Bay port of Alviso some ten miles north where he put them on a boat for San Francisco. There they were transferred to a Sacramento River steamer and arrived the next day in Sacramento near 70 miles upstream. Mr. Buck left his bees in the Sacramento area for three months to gather the summer nectar flow from the plants watered by the overflow of the Sacramento River. In the fall he moved them back to his home apiary. (Watkins 1968)

1905 On the Road

Throughout the bee magazines of the time, there is frequent mention of "Migratory Graham," who seemed to think nothing of putting his hives on the road, whenever greener pastures appeared on the horizon:

A California bee-raiser moves his hives, during the season to get the best pasture, and secures thus three crops of honey. The orchards of the Sacramento and San Joaquin valleys burst into bloom some months before the southern sagebrush, and to them C. I. Graham drove a wagon-load of bee colonies late in January, 1896. As the blossoms faded in April, Mr. Graham turned southward again. It was predicted that his bees would desert him by the wayside, for, while

> bees can be moved with impunity during their quiescent winter, it
> is generally considered impossible to transport them during their
> active season. (Friends 1901)

According to the article in the Philadelphia publication *Friends' Intelligencer and Journal*, Mr. Graham had hit upon the tactics that would later become standard practice. He would move into the Sacramento and San Joaquin valleys in January to assist the pollination of fruit trees. This would be followed by the pursuit of a honey crop from the wild sage that still grows abundantly along coastal California. Finally, he would move into irrigated alfalfa, a practice that continues today. One grower was so pleased by the results of bee pollination that he offered Graham a permanent home for the hives and a share of the profits from the sale of the hay.

The Automotive Age

The idea of motorizing carriages and wagons began in the 1800s using everything from steam power and electric motors to gasoline engines. The term "car" had been in use for centuries, and implied a vehicle pulled by a horse. An automobile was therefore, a car equipped with its own power. This was soon shortened to "auto" and larger vehicles were referred to as "auto-trucks." One of the most prolific writers about beekeeping in the late 1800s was John Martin, better known as "The Rambler." In his Ramble of January 1, 1900, he made the following pronouncement.

> The automobile will revolutionize the bee-keeping industry. The
> automobile will supersede the horse in the moving of things to and
> from the apiary, and everything in and around the apiary. We can
> even fix to move bees rapidly to out-locations when desirable, also
> to move honey to market. (Rambler 1900).

The Rambler had quite a following and he wrote well over 200 articles. In the beginning they were about brief visits to his neighbors in his home state of New York. Eventually, he went to California and pretty much told the whole beekeeping world what was going on there. The state had been swarmed by prospectors hoping to cash in on the Gold Rush. This was accompanied by many more people who had in mind selling the miners everything from coffee and sugar to Levis and top hats. But the real boom came when it was realized that California was blessed with an abundance of sun and rain; it was destined to become in turns the bread basket, the dairy farm, and finally the fruit orchard of the nation.

Beekeeping followed all of these developments and the Rambler chronicled the situation from by whom, when and where the huge honey crops were made. Much of what he wrote in 1900 still applies. Beekeeping began around Sacramento with John Harbison leading the pack. Honey was gathered from plentiful wildflowers, but the prospects came and went with the flooding of the great valley. Eventually dams were constructed in the foothills to capture the snow melt, and the big valley was turned into a giant farm. As the Rambler noted, prospects for beekeeping were far better in the southwestern part of the state, where wild sage and buckwheat blanket the hills. While trucking bees may have seemed like a boon to some, others had mixed opinions about the prospect of bee hives being moved all over the country. The following appeared in the magazine *Gleanings in Bee Culture*, in June 1912.

> *Migratory beekeeping has not been practiced to any extent in the United States. One difficulty is that our distances are so great, and the freight is so high, that most beekeepers lack the nerve to try it. Quite a few beekeepers in Northern Michigan have managed, however, to catch their clover and basswood, and then move north ward by car to catch raspberry and fireweed. It is my opinion that migratory bee keeping might be practiced to more advantage than it has been. Just now the beekeepers in some western States are moving their bees by the carload, after they secure one crop, into California to catch the alfalfa or sage that comes on later. At this writing, May 27, it is stirring up a hornets' nest, or, rather, a bees' nest, among the California beekeepers, who fear the introduction of disease, and who complain that their territory is already overstocked. The fight is on, and how it will be settled remains to be seen.*
> *(Gleanings 1912)*

Bees on the Road

The bee journals were quick to post stories about the new mode of transport. J. E. Pleasants of Orange, California wrote this in the *American Bee Journal*, in September 1915.

> *Perhaps no State in the Union uses more automobiles than California. In fact, some think we run to extremes here in that line. Auto trucks are now used a great deal for moving both honey and bees. While I am a great lover of the horse, and do not like to see them altogether replaced by machines for pleasure driving or even*

draft work, it seems to me that this is a line of work in which the machine especially shines. The distances are usually great from out- apiaries to market, also the rapidity with which bees can be moved from one locality to another is of course a great advantage. (Pleasants 1915)

MOVING FOUR WAGONLOADS OF BEES BY TRACTION-ENGINE IN CANADA.

1910 Moving four wagonloads of bees by traction-engine in Canada

In 1916, Max Clemens Richter wrote his master thesis on the subject of "Beekeeping in California," possibly the first of many such research projects.

Migratory beekeeping in California is a well-established business, and with the improved methods now in practice in preparing bees for moving, the California beekeeper shows no hesitancy in transporting a strong colony during the summer time, on even a journey of considerable length. (Richter 1916)

In his 1919 book *Success with Outapiaries*, M. G. Dadant gives several examples of the utility of trucks in extending the range of their "outapiaries" by migrating within a limited distance:

But with the elasticity in opportunity for outapiary expansion by means of the automobile and truck, he should do more than this; he should study carefully his territory for 100 miles in each direction. He may, by this same means, place his apiaries to best advantage, and he may, moreover, change locations (migrate) with his bees to an extra crop. An instance of this may be mentioned in the case of the Dadant apiaries during the season just passed. White clover, our main flow, was a failure, and such little as there was, together with sweet clover, was used in making increase. Careful observation showed us, however, that the drought had not affected the growth of weeds in the Mississippi bottoms some distance away. By the aid of two large trucks all of the 700 colonies in these apiaries were moved into the bottom for the added harvest with the result that a haul of forty miles at the most meant an average from 70 to 100 pounds to dearth, with consequent necessity for liberal feeding to get colonies in condition for the winter. (Dadant 1919)

1914 auto truck adapted to the business of honey production

Eventually, the use of gas powered trucks caught and became the norm for beekeepers from coast to coast. By 1920, J. E. Crane of Middlebury, Vermont, described the transition from the old mode to the new.

In conducting outyards an auto-truck or at least an automobile is a great help. I used horses for more than thirty years without any

serious accident, but there is always danger where horses are used
unless great care is exercised. It is always well if horses are used
to have the road descend or down grade from the yard so a load
of honey can be run by hand for some distance from the yard if
necessary before horses are hitched to the wagon. There are other
and very important advantages in the use of an auto truck.
(Crane 1920)

Fine Tuning the Practice

Lee Watkins was born in 1908, in the town of Selma which is situated in the heart of the San Joaquin valley. His father was a beekeeper and so he was immersed in California beekeeping from a young age. Watkins went on to write about its history and the practices of early to mid-twentieth century beekeepers. He attended the University of California at Berkeley where he gained an appreciation of anthropology from people such as the famed Alfred Kroeber, whose work on the native Californians is still the best source material. Eventually, he worked at the apiary at U. C. Davis, which is one of the world's prominent research facilities today. Watkins wove together such disparate subjects as beekeeping, philosophy and the art of tying of the "truck hauler's hitch".

Perhaps most notably, Lee wrote about his experience hauling bees around the foothills of the Big Valley. The custom by then was to move bees five or six times a year, from the irrigated farmlands to the surrounding low mountains, depending on how much rain fell and when. He and the crew laboriously fastened screens to the front and tops of the hives in an effort to confine the bees and not lose the valuable field force.

One time they were tasked with moving several hundred hives down rough mountainous roads in the summer where the temperature was going up to 106°F. In the effort to get the job done swiftly, they loaded the bees late in the day and moved them. While the bees came out all over the hives, they did not fly off. To their great surprise and pleasure, no bees suffocated during the move, as they often did when screened in. The agitated bees crowd the screens like a stampeding crowd and block the airflow, causing the bees inside to overheat and smother. After this successful move, they abandoned the use of screens. Lee wrote about his success in the *American Bee Journal*. He said: "The practice has gradually increased until now almost all of the beekeepers in California move bees without screens".

A Load of Twin-Mating Nuclei Hauled by a Sting Proof Horse from one Mating-Yard to Another.

1915 auto-truck revolutionizes beekeeping

A final word from Lee Watkins. Sadly, he died in 1972, not quite 64 and two years before I got into the bee business myself. In 1950, he had the foresight to make the following prediction:

> *When the beekeeper is no longer the forgotten man of agriculture and he gets paid a fraction of "twenty times greater value than the honey and wax the bees produce [the estimated value of pollination]," beekeeping practice is going to change in a great many sections of these United States. More beekeepers than ever will be moving their bees two or more times a year and will make as much or more money by furnishing pollination service with strong colonies of bees than they will produce from honey crops. (Watkins 1950)*

While it's true that trucks made it possible to carry on beekeeping at a much larger scale, they didn't necessarily make the work easier; perhaps just the opposite. Where once thirty hives was considered a big load, trucks were soon hauling loads of 120 or more. These would be laboriously loaded by hand, sometimes in hot and humid weather, occasionally accompanied by vehicle breakdowns, resulting in unforgettable episodes of stinging. I will continue this story in the next chapter, in which the mechanical hive hoist makes its appearance in the late 1940s.

Works Cited

Adams, R. L., & Todd, F. E. (1939). *Cost of producing extracted honey in California* (No. 656). US Department of Agriculture.

Biggle, J. (1909). *Biggle Bee Book: A Swarm of Facts on Practical Bee-keeping, Carefully Hived*. W. Atkinson Company.

Crane, J. E. (1920). Outyard Yards. *Domestic Beekeeper*, Vol. 33, No. 4.

Dadant, M. G. (1919). *Outapiaries and their management. American Bee Journal*.

Eckert, J. E. (1954). *A handbook on beekeeping in California* (Vol. 15). California Agricultural Experiment Station Extension Service.

Friends' Intelligencer and Journal. (1901) Vol. 58, No. 25

Gleanings in Bee Culture. (1890) Vol. 18, No. 8

Gleanings in Bee Culture. (1912) Vol. 40, No. 12

Pleasants, J. E. (1915) American Bee Journal. Vol. 55, No. 9

Rambler (John Martin). (1900). Ramble 180. *Gleanings in Bee Culture*. Vol. 28, No. 1.

Richter, M. C. (1911). *Honey plants of California*. University of California Publications.

Richter, M. C. (1916). *Beekeeping in California*. University of California.

Watkins, L. H. (1939). Moving Bees Without Screens. *American Bee Journal*. Vol. 79, No. 11.

Watkins, L. H. (1950). Pollination service and moving bees. *Gleanings in Bee Culture*. Vol. 78, No. 6.

Watkins, L. H. (1966). The Truck Hauler's Hitch: A Boon to Migratory Beekeepers. *Gleanings in Bee Culture*. Vol. 94, No. 9

Watkins, L. H. (1968). An interesting report on the history of migratory beekeeping in California. *Gleanings in Bee Culture*. Vol. 96, No. 12.

Zierer, C. M. (1932). Migratory beekeepers of southern California. *Geographical Review*, 22(2), 260-269.

Chapter 16

Modern Beekeeping

People have been wrestling with the weight of bee hives for centuries, but the modern era of beekeeping brought with it high numbers of hives which tend to be much heavier due to better management. A honey filled straw skep might weigh a hundred pounds but a three story hive could weigh double that.

Although electric hive loaders appeared as early as the 1950s, they were particularly fussy and required skill to operate properly. Many beekeepers chose not to adopt them and continued to move hives the conventional way, by hand. I had firsthand experience in this type of migratory beekeeping which persisted well into the 1970s and 80s. The colonies were typically housed in two-story hives, which would be given a hearty dose of smoke and lifted up to the truck bed by two strong people.

Once on the bed, the hives would be double stacked and roped down sturdily employing the trucker's hitch. Nets were nowhere to be seen. Ideally the hives would be loaded at dusk and moved in the dark, to be unloaded at daybreak. Sometimes the work occurred under less than ideal circumstances. When handling bees at night, they switch from flying to crawling; they quickly wind up inside your clothes if there are openings. We were often stung hundreds of times. Some commercial beekeepers still rely on cheap inexperienced help to do the work that others have machines to do.

Eventually, forklifts became the norm and large trucks are now seen hauling 500 or more bee hives between the north and south and from coast to coast. About half the colonies of bees in the United States are moved to California to pollinate almond blossoms in February and March. From there they fan out to apple orchards, cranberry bogs and blueberry patches as far away as Maine.

Mechanical Aids

As early as 1907, people were actively contemplating the use of mechanical devices to assist with heavy lifting. The following excerpt is from *The Bee-Keepers' Review* which began publication in 1888 in Flint, Michigan.

A Hive Lifter ought to be the next apicultural invention. Several of these contrivances have been put into use, but just how practical they are I can't say from actual experience. The most of them are of the "stump puller" type, a tripod of light, yet strong timber, with some sort of a contrivance like a set of pulleys or levers for lifting the hive. E. M. Hayes, of Veedum, Wis., suggests the use of a tackle-block wire stretcher, such as is used in stretching wire when putting up wire fence. A hive lifter ought to be light, easily and quickly set up and attached to the hive, and allow the operator to lift the hives easily and with the least possible expenditure of time. Inventors, go to work on the problem. (Bee-Keepers' Review 1907)

1908 Mrs. Anderson's Hive Lifter

Not everyone was keen on mechanizing, however. In 1908, Mrs. J. L. Anderson writing in the *American Bee Journal* described "The Old Reliable hive-lifter: self-adjusting, easily controlled, such as every lady bee-keeper should have." This was, of course, her husband. She noted, it was self-adjusting to fit any size or shape of hive. It was "easily controlled" and could also be used for lifting carpets and other tasks. She admits it was expensive but in the long run "the best is often the cheapest." Hers had been in use for many years and was "almost as good as new."

Woven throughout the history of honey production is the story of inventors. A. I. Root certainly is an example. He championed many mechanical devices from his wind powered printing press, on to his keen interest in the Wright Brothers and their flying machines. The Root Company offered motorized honey extractors in a range of sizes. He probably would have invented a hive loader for trucks, had he not died in 1923. The Illinois State Beekeeper's 1912 report described a "practical hive lifter" with which one person could raise heavy boxes. Built out of iron pipe, it had a cross arm with a track for a pulley to run back and forth. Attached to the pulley was a "camp chair" which would grab the hive which could then be raised by "pulling lightly."

HONEY-LOADER.

Having a considerable amount of honey to load, the task was lightened by a mechanical loader, an upright carrying a cross-beam at one end, with rope and grab-hook, and at the other end a trail-rope. This readily lifted the 140-lb. cases from the ground to the wagon.

1915 Easy Honey-Loader

Small hoists powered by compressed air or electric motors were becoming common in factories and machine shops by the 1920s. An article in the *Industrial Engineer* magazine described the features of electrical hoists:

> *The modern successor to the chain block, is the electric hoist in ½-ton to 2-ton capacities. While the factors which justify the larger investment may be quite varied, there is always one underlying reason for the choice of the electric hoist. This is the human factor; man power is multiplied by its use and men can be selected on a basis other than brawn. For this reason, in selecting a hoist, careful consideration should be given to all the parts that make up the*

hoist, and in particular to the control, which is the link between the operator and the hoist, and between the hoist and the source of power. (Industrial Engineer 1924)

Migratory Beekeeping, California Style

Anyone who is interested in the history of migratory beekeeping will find a bounty in the writings of Lee Watkins and Clifford Zierer. The former was an apiculturist at UC Davis and wrote extensively on the history of California beekeepin, as mentioned in the previous chapter. The latter was a professor of Geography at UCLA but his writings reflect a strong interest in mechanization and its effect on modern agriculture. His writing ranged from "Geography and the Automobile" (1922), "The Fishing Industry of California" (1935), and "Migratory Beekeepers of Southern California," (1932). Zierer was well acquainted with these topics and clearly described the progress of beekeeping from the days of horse drawn wagons to the trucking era. Underlying all of this was the need to move about, to make beekeeping pay. Whether to escape droughty conditions which parched bee forage for years at a time, to the lure of greener pastures just over the mountain range, the goal was the same. According to Zierer, beekeepers were using the railroads to move their hives from the California coastal ranges, inland to the San Joaquin/Sacramento Valley to reach vast alfalfa fields. By 1932, California was the leading commercial beekeeping state. It claimed 400,000 hives run by about 10,000 beekeepers. He estimated that of these, some 2000 were migratory beekeepers, moving bees to benefit from the variety of honey plants. Zierer pointed out that while 150 plants had been described as important to honey production, at least locally, only eight were major players. These were alfalfa, black sage, orange blossom, purple sage, wild buckwheat, white sage, lima beans, and star thistle.

When I lived in southern California in the 1970s and 80s, lima beans were long gone but eucalyptus had replaced them as a major source of honey. Zierer noted that even then in the 1930s, the Blue Gum eucalyptus was common from San Diego clear to San Francisco and yielded nectar and pollen in winter, providing an important source of food for colonies to build up strength. Already, nut and fruit growers were paying from $1.50 to $3.00 per hive to have bees brought in for a few weeks to pollinate everything from pears, almonds, apples, and cherries. On the other hand, beekeepers were known to pay $15 to $30 for locations where they could get a good crop of orange blossom honey. (To get a sense of these dollar values today, multiply by about 10).

The Boom Loader for Hives

One of California's earliest bee champions, and certainly the most well-known, was John Edward Eckert (1895-1975). Shortly after arriving at the University of California at Davis in 1931, he was an assistant professor of Entomology and assistant apiculturist. Professor Eckert served as chairman of the Davis entomologists from 1934 to 1946. Eckert understood that honey bees were an integral part of California agriculture. Starting in 1949, he wrote a series of articles for the various beekeeping magazines on the topic of hive loaders. In his words:

> *Beekeeping is an industry in which a certain amount of heavy lifting can be considered as an occupational hazard. The hive measurements are such that one cannot avoid an awkward stretch of the shoulder and back muscles while lifting a hive or heavy super which frequently leads to pulled muscles or to a displacement of a vertebra or of the sacroiliac. The loading and unloading of hives of bees or of heavy supers of honey on uneven ground is probably the hardest part of beekeeping. (Eckert 1949)*

Eckert credits A. K. Whidden of Colton, California, with devising a rope and pulley system for lifting hives during World War II. While it was fairly complicated and slow, it enabled him to manage over a thousand hives by himself, at a time when there was an acute shortage of laborers. This arrangement incorporated the boom which could reach hives on either side of the truck and a carriage with forks that could grab the hives under the hand cleats. According to Eckert, Dan Aten, the Dyer Brothers, Allred Brothers, Chas. Reed, Harold Sharp, W. D. Holton, Jack Shackleford, Oliver Hill and other California beekeepers designed different types of hive loaders. Oliver Hill and his father were located in Willows, CA, and maintained more than 2000 colonies. They named their boom loader "George," as he was a valuable member of the crew when it came to moving hives and heavy supers filled with honey.

The Hill's design was unique, they made it themselves from available parts. The motor for raising the load was not mounted at the front of the boom like most later devices, but was directly above the carrier, and moved back and forth on the repurposed overhead garage door track. The hoist was driven by an electric truck starter motor powered by a "third rail" on the boom track that carried electric current from a battery to the motor. The arrangement required the bare minimum of cable; later boom loaders ran the cable the whole length of the track. Using the track to transmit the electricity minimized the need for dangling electric cords. Since the loaders were made by individual beekeepers, they varied considerably in their designs. Eckert described a more sophisticated and powerful loader:

> It was developed by the Valley Pollination Service of Orange, California, and engineered by the Hopper Engineering Company of Bakersfield with the financial aid of the Maricopa Seed Farms and beekeepers. It is operated by two motors powered by a 2½ H.P. generator housed in the enclosed portion extending over the cab of the truck. Push button controls are mounted on the handles of the fork lift to fully regulate the movement up and down, and along the boom which is 16 to 20 feet in length. (Eckert 1949)

This design has the cable running the length of the track, down to the carriage. This means the operator can use power not only to lift or lower the hives, but also to move them forward and backwards along the boom. Most of the early devices required a lot of skill and strength to maneuver the loads, especially since they had a tendency to swing away at times from the operator.

Another approach was the use of hydraulics. The Dyer Brothers used the hydraulics from the truck transmission to power a motor at the bottom of the boom support. The boom was mounted in the middle of the truck bed and capable of swinging a full circle reaching either side and in back of the parked truck. It was much more heavy duty and could lift 1000 lbs, more than the weight of a full barrel of honey. When not in use, it was lowered into a cradle over the cab. Hive loaders were also equipped with bright lights to facilitate working at night.

Beekeeping for Women and Solitary Men

Dr. Eckert certainly regarded the boom loader as one of the most significant advances in modern beekeeping. He wrote many articles and profusely illustrated them with clear photographs. In 1955, he wrote a piece titled "Beekeeping for Women," which shows Miss Dorothy Ryder wearing a plain house dress, lifting three full supers of honey. Eckert noted that California had fewer women beekeepers than other states, due to the migratory nature of beekeeping in the state. But, he assured his readers, "This situation is likely to change in the future, due to the mechanical aids now available."

Some of our older beekeepers who are now using the hive loader state, with smiles, that this mechanical aid has taken the backache out of beekeeping and has extended their active work with bees by many years. Even the younger beekeepers say they would not be without its services, now that they have become accustomed to moving bees simply by the push button controls. (Eckert 1949)

BEEKEEPING FOR WOMEN NOW MADE POSSIBLE BY THE HIVE LOADER

by Dr. J. E. Eckert, Davis, California

1955 Dorothy Ryder, demonstrating hive loader

Not all of these devices were sophisticated heavy duty affairs. In 1953, the *American Bee Journal* ran an article titled "A One-Man Hive Hoist" by Chas. S. Hofmann. Charles was working "in the 300-400 colony range which, of necessity, must now be termed strictly a one-man outfit (a one-man outfit is where there is never anyone to take hold of the other end of anything)." He decided to sink $200 into a hydraulic "mast and boom." This incorporated an electric motor run off the truck battery which in turn powered the hydraulic pump which raised and lowered the boom. The load was raised by a push button switch on the carrier; he had a release valve to lower the boom. The whole unit was about 175 pounds and could lift up to 200 pounds. Compare this to the boom loaders, which often added 1000 pounds or more to the weight of the truck. In his article Hofmann says that over the first two seasons he moved some 350 hives twice and about 110,000 pounds of honey. "The tongs never damage or even mark a hive, and the hoist handles colonies of bees and supers of honey more gently than in most cases can be done by hand," he wrote.

Walter T. Kelley and Modern Beekeeping

According to Prof. Roger Morse, Walter T. Kelley (1897-1986) was "one of the more colorful people in the beekeeping industry." Walter Kelley began producing package bees and queens in 1924. Soon he was making and selling beekeeping equipment. He moved to Kentucky in 1935 and renamed the company: The Walter T. Kelley Company. His supply catalog was the largest in the industry, and he relied on direct sales to beekeepers. Kelley also wrote extensively about apiculture and from 1947 to 1956 he was the editor and publisher of his own bee journal, *Modern Beekeeping*. Beginning in 1952, the magazine featured the spate of articles by J. E. Eckert about the various boom loaders that beekeepers were developing in California. These were profusely illustrated with Eckert's clear photographs. Soon, Kelley patented a version of his own and began selling them to beekeepers around the country. An ad from 1966 proclaims:

> Beekeepers tell us that we build the best hive loader made and our sales prove it. We have sold hive loaders in Australia, New Zealand, France, Canada and over much of the U.S. Our machines operate on 12 volt systems using one battery plus the truck battery. No need to run the truck engine while operating. Push buttons on the cradle actuate motors to level the boom cross ways and length ways of the truck. The loader does the lifting and moving once you learn how to use it. Will lift 5 hive bodies full honey [about 400#]. (Kelley 1991)

By the end of the 1960s, commercial beekeeping had already expanded to a grand scale. Beehives were in demand for pollinating vast acres of fruit and nuts, especially the swaths of almonds that were being planted in California's Great Valley. Earl D. Anderson wrote up a comprehensive *Appraisal of the Beekeeping Industry*, published in 1969 by the U.S. Department of Agriculture. Among other things, he described the scale of the undertaking:

> Some of the large migrant beekeepers use diesel-powered semi-tractor trucks with another trailer behind for long-distance hauls.

The cabs of these units may have sleeping accommodations also. One beekeeper in the East reported using common carrier trucks to transport his bees to another State for pollination. He rented a forklift truck for loading the transport truck at his yards. Then the growers at destination placed the colonies for pollination. (Anderson 1969)

KELLEY'S HIVE LOADER

1966 Kelley's Hive Loader

This reads like something we might see today, in one of the many news articles where the reporter has "discovered" the fact that a few million hives hit the road every year. They are liable to be in California in February, Florida in March, New York state in May, and on up to the blueberries in Maine. The fact that the price paid for pollinating almonds is so high, it attracts beekeepers from the length and breadth of the USA. In a way this has been a good thing for all beekeepers. Having large scale operators concentrate more on pollination and less on honey keeps the price of honey strong, especially for small holders who provide "local honey" to their regional customers. Clearly, if growers stopped paying so much for pollinating units, large scale beekeepers could switch back to producing more honey and the prices paid for honey would no doubt plummet.

Works Cited

Anderson, E. D. (1969). *Appraisal of the Beekeeping Industry.*
US Department of Agriculture.

Bee-Keepers' Review. (1907). Published in Flint, Michigan.

Eckert, J. E. (1949). "Mechanical Aids to Moving Hives and Supers."
In: *Report of the State Apiarist.* Published by the State of Iowa.

Hoffman, C. S. (1953). "A One-Man Hive Hoist." *American Bee Journal.* Vol. 93,
No. 4. p 148-49.

Illinois State Beekeeper's Report. (1912). The 22d Annual Meeting of the Illinois
State Bee- Keepers' Association. Springfield, Illinois, October 30 and 31, 1912.

Industrial Engineer. (1924). McGraw-Hill Company, Chicago. November, 1924.

Kelley, W. T. (1991). *How to Keep Bees and Sell Honey.* 12th Edition.
Walter T. Kelley Co. Clarkson, Kentucky.

Modern beekeeping. (1946-56). Published by Walter T. Kelley. Clarkson, Kentucky.

Morse, R. A., Flottum, K., & Root, A. I. (1990). *The ABC & XYZ of Bee Culture: An
Encyclopedia Pertaining to Scientific and Practical Culture of Bees.*
AI Root Company.

Watkins, L. H. (1968)."Migratory Beekeeping In California."
Gleanings In Bee Culture. Vol. 96, No.12. p 732-34.

Zierer, C. M. (1932). "Migratory beekeepers of southern California."
Geographical Review. Vol. 22, No. 2. p 260-269.

Chapter 17

Pioneers of Texas Beekeeping

How honey bees got to California is fully documented, partly because it was such a difficult undertaking to get live bees from New York to California, via Panama and over land. On the other hand, there are many recorded observations of honey bees being "already there" when settlers moved west into states like Kentucky, Texas, or even New York state. Native Americans realized settlers were getting close when swarms of bees arrived in the woods where they hunted; the heralding of a change to come. So this it seems, is how the honey bees got to Texas: they migrated ahead of the westward advance of the white men.

Honey Trees

Mary Austin Holley was one of the most prolific and respected chroniclers of Texas history. Her writing provides us with a rich and detailed story of the early days. In the autumn of 1831, she penned a series of letters while visiting the settlement of Stephen F. Austin, which had been established some ten years before, in the location of what is now the city of Austin, Texas. At that time, the territory was part of Mexico which granted him permission to settle 300 families there and another 1700 were permitted by 1831. In one of her letters, Holley describes honey trees.

> *Upon inquiring what kind of tree was meant by honey tree, I learned, that hollow trees, in which the bees deposit their honey, are so called. These trees are very abundant, and honey of excellent quality and in any quantity, may be obtained from them. These trees are also called bee trees. There are persons here, who have a peculiar tact in coursing the bee, and of thus discovering these deposits of the luscious store. This employment is not a mere pastime, but is profitable. The wax alone, thus obtained, is a valuable article of commerce in Mexico, and commands a high price. There is always a demand for it, it is so much used in the churches. This, it will be remembered, is a Catholic country. In some of the churches, the wax candles made use of, are as large as a man's arm. It often happens, that the hunters throw away the honey, and save only the wax. (Holley 1833)*

About the same time as Mary Holley was writing her travelogue, Hugh Kerr composed an epic poem which depicted the "battles, rivers, lakes, streams and animals of the country." His book of poems was the first ever produced in the fledgling state. Born in Ireland, Kerr came to America around 1795. He moved to Texas in 1832, and died there some ten years later. This is a section of his poem which applies to this story:

> If shady groves attract you there,
>   The honey bee comes flitting by,
> Well laden from the prairies, where
>   It soon collects a sweet supply.
> Some hollow tree its load receives,
>   And hundreds follow in its train,
> Though often robbed by cunning thieves,
>   Their storehouse is supplied again. (Kerr 1838)

In a few short lines, he corroborates the idea that honey bees were already abundant and hunted for their treasure. Probably, the keeping of bees in log hives commenced about the same time. It made sense to cut a bee tree down and bring it home, where it could be pilfered regularly without much trouble.

The Beginnings

The beginning of beekeeping in Texas is chronicled in two specific works, which it was my good fortune to be able to borrow from the libraries that held them. These are both master's theses; the *Origin and development of the bee industry in Texas*, was written in 1938 by Thelma May Burleson whose father was a pioneer Texas beekeeper. *Apiculture in Early Texas* was completed in 1952, by Clark Dumas.

Sadly, these types of works are tied up by copyright laws and cannot be read by the general public. Further, the holders of the rights have long since passed on and cannot release the works into the public domain. They also contain much material that has been quoted from earlier books and magazines, many of which are even harder to track down. Writing about beekeeping in Texas prior to 1872, Burleson flatly states that it "was of so little consequence that as an industry it hardly merits discussion further than to point out that it did exist." This was at a time when honey production was well under way in many other states, including recently settled California.

Wilhelm Bruckisch

Wilhelm Bruckisch

Just as the beginnings of beekeeping in other states can be tied to specific individuals such as Langstroth in Pennsylvania, Root in Ohio, Harbison in California, etc., so can we point to them in Texas. One such was Wilhelm Bruckisch, who was born in 1802 in Postelwitz, Silesia - what is now southern Germany, very close to the Czech border. Bruckisch was a friend and associate of Johann Dzierzon, the master of beekeeping in Europe. Dzierzon was also from Silesia, the part which is now Poland, but he wrote and spoke in German. When Bruckisch moved to Texas in 1853, with his wife and five children, he began keeping bees using the techniques taught by Dzierzon. Bruckisch settled in Hortontown, which is now New Braunfels, just northeast of San Antonio, where he died in 1877. Bruckisch's sons fought and died for the Confederacy in the Civil War; their daughters married German Texans.

Bruckisch was a prolific writer in German and English. His *Bienenbuch* (*Bee Book*) went through six editions from 1847 to 1861, and *Besste Bienenzucht-methode nach Pfarrer Dzierzon* (*Best beekeeping method according to Pastor Dzierzon*) came out in 1867. But he was most well known in the United States for his entry in the 1861 Annual Report of the US Patent Office. This was an extended piece on "Bee Culture" which showed not only his advanced understanding of the subject but his close familiarity with Texas beekeeping and what made it different from elsewhere.

> *Many a hive with us (in Texas) containing but a handful of bees in the month of August requires only sufficient rain to become populous again within three weeks. In another three weeks every room and space of the hive will be filled up, while if it should enjoy favorable weather for two or three weeks more, every space that has been emptied will be again filled with a second crop. Generally speaking, the South is far more adapted to the bee than the North. Though an oppressive heat is not favorable to it, even if it rest a little during the warmest hours of the day, yet cold is much more injurious, as it causes death. While it is well known that it can live at the equator, it is not yet ascertained how far North it may do so. (Bruckisch 1861)*

In the 1860 report, Bruckisch advocated state and federal research on bee culture problems and the formation of a national beekeeper's society. He asserted that a beekeeper, with help, could manage one thousand hives. He added that a square mile could sustain 200 to 500 of them when conditions are favorable, but that none would do well in bad weather. Bruckisch recommended proper fencing and dogs to repel cattle and thieves from the apiary. Despite his knowledge and experience, he kept and advocated the Dzierzon hive at a time when the Roots and Dadants were already selling many thousands of hives based on the model which Langstroth introduced to the world in 1853. The Dzierzon hive was a sort of tall cabinet, which opened from the side to access the combs. John Harbison also promoted this type of hive all over California, during the 1860s. These cabinet hives fell out of fashion by the end of the century, superseded by the more practical frame hive which was accompanied by the use of the honey extractor. Bruckisch seemed to be aware of the other hives, since he wrote:

> The omission to mention in the course of this article the good points
> of other hives was not from a want of attention to them, but because
> the Dzierzon hive combines all the good points of a useful hive, and
> from a conviction it will contribute largely to the general promotion
> of bee-culture to urge a uniform adoption of his method and hive.
> (Bruckisch 1861)

According to Thelma Burleson, Bruckisch's 200 page book written in German contained the entire system developed by Dzierzon, details of the hive, and "sandwiched in between the paragraphs … notes telling of his experience in Southwest Texas." She wrote that these anecdotes would still be of interest to the beekeepers of the time, i.e., the 1930s. My research concludes that while it was translated into English, only the German version was actually published. According to Burleson, Bruckisch was "probably the first person in Texas to become interested in the commercial possibilities of bees."

Jennie Atchley

It would be impossible to write a story of pioneer beekeeping in Texas without Jennie Atchley. In her own words, written in 1904:

> Thanks for your kind mention, and I will state that when I came to
> Texas in 1876 there were only five practical apiarists in the state,
> and to look back at the past and then view the present proportions
> of apiculture, it hardly seems that such strides could be made in so

short a time. I mailed the first queens sent out of the state, and sent out the first pound of bees by express, so far as I know, and have written enough on bee culture to reach many miles, if placed in a line. My bee-keeping history in Texas would fill quite a large book. (Atchley 1895-1904)

As mentioned, one of my primary sources is the 1952 Dumas thesis. He quoted extensively from the *Texas Farm & Ranch Magazine*, which is a good thing because the journal is nearly impossible to obtain anymore. Evidently, Jennie Atchley was introduced to the readership of this publication in 1893 like this:

"The Greatest Bee Family:" There lives in Beeville, Bee County, Texas, Mrs. Jennie Atchley, the most extensive breeder of bees in the world. She is 36 years old and the mother of eight children who with her do all the work in her large apiary. (Dumas 1952)

Jennie was born in Tennessee in 1857. She married E. J. Atchley in 1876 and they moved to Dallas County, Texas. They moved to Beeville in 1893, and established a farm of 120 acres on which they raised "everything needed for the family." According to *Farm & Ranch*, Jennie and E. J. were selling about 5000 queens a year, running 1000 production hives and another 1000 queen mating hives, plus two to five freight carloads of honey. The Atchleys considered horsemint (*Monarda punctata*) to be a primary nectar source, as well as cotton which blooms in July. But they were also well acquainted with the various regions of Texas, especially the southwest centered around Uvalde where cat's claw (*Acacia greggii*), and mesquite (*Prosopis glandulosa*) are common. She told her readers that they would likely get some honey anywhere in Texas but "out in the hills of Southwest Texas is the best place."

I have included a photograph of the whole Atchley family that appeared in 1893, with an article describing all of the members and their various roles in the family business. On the left is Willie, the oldest boy, whose main work was queen rearing and all of its esoteric practices. Charlie's job was carpentry; next is six year old Leah, who is a runabout helper in the bee yard. Little Ives appears to be chewing on a frame of honey. Mr. and Mrs. Atchley stand tall, each with their respective favorite books: *The ABC* by Root, and *Doolittle on Queen Rearing*. The tall girl is Amanda, "one of the best beekeepers I have," wrote Mrs. Atchley. Far to the right stands Napoleon; she said of him: "He is going to be a bee-man some day, if he lives, notwithstanding he loves chickens." The "last and least" is the baby Thomas.

Atchley Family

Mrs. Atchley became a regular contributor to the bee journals of the day, answering questions from all over with patience and thoughtfulness. Her way with words was not unnoticed:

> I have been repeatedly urged to start a bee-paper but after due consideration, I concluded that it was best and safest not to do so. Having already been permitted "to ride" a little way upon "journalistic waters," I find that many times "the sea" is rough. Therefore, I have made arrangements to ride in one of the old, reliable, trustworthy and well-tried "boats"—the American Bee Journal—and I shall feel much safer there, than in a "new boat" of our own.

As it turned out, the Atchleys did start a "bee paper," which they named the "Southland Queen." They also branched out in other directions, selling beekeeping supplies, equipment, and bulk bees. She continued to write in her colorful way:

Oh, I must tell you of a wreck I had in moving my bees. I had moved all on wagons except my fine breeders, and with the last loads I put said bees in my family hack and took them behind my wagon. It was a nice warm day Dec. 31st and being scarce of hands my daughter, Amanda, drove one wagon, a fine large team of mules, one of our thresher teams. Coming through a large pasture out of one into another through a gate a herd of mules and horses ran through the gate just behind Amanda's wagon and scared her team. They ran away and struck the hack and capsized it, bees and all. Amanda jumped off and the mules ran on and Willie galloped ahead of them. Strange to say none of the queens got killed, but one hive lost many bees, and not a bee offered to sting anybody or our teams.
They were of the yellowest type of the five banded Italians.
We straightened all up and pursued our journey; nobody hurt.
(Atchley 1892)

Atchley Apiary

Their magazine ran from 1895 until 1904 when the Atchleys moved again, to California. I haven't been able to learn what they did there, as there is no further writing from or about the prodigious Atchley clan. I like to think Jennie Atchley made a bundle in the bee business and retired to California to live the rest of her life by the beach.

Sallie Sherman

Because Mrs. Sherman's story was so remarkable, it was well documented. The *Texas Farm and Ranch Magazine* ran a comprehensive biography about her in 1889. She was born Sarah Elizabeth Johnson in Decatur, Georgia in 1843, being one of seven children. Her parents, the Rev. Thomas C. Johnson and his wife Abigail, moved the family to Texas in 1856. According to the magazine, they arrived in "almost destitute condition from fires, storms and sickness," although they don't elaborate on these calamities, perhaps due to their being commonplace. The family learned a variety of skills from farming grapes and tobacco to making wine and cigars. The article says young Sarah "made as many as one thousand a day, yet never learned to use the noxious weed."

Mrs. S. E. Sherman

In 1859, Rev. Johnson bought 160 acres of farmland in Burleson county, not far from College Station, Texas. At the same time, Mr. S. G. Sherman settled nearby, and he and Sallie, as she was called, became engaged in 1862. It was not until 1866 that they were married, soon after the end of the Civil War. By then Mrs. Sherman had acquired many more skills including knitting, weaving and making palmetto hats. Being adept at so many things no doubt was a great help, as her husband died in 1868, leaving her with their young son. Sallie was left with a large tract of land, a few horses, cattle, and hogs. According to the magazine, her health had always been poor, but she "took to outdoor life." She was determined that anything dependent upon her would never lack proper attention.

Mrs. Sherman moved to the town of Salado, Texas in 1875, in order that her son should receive a proper education. Not too long after that she became interested in bees. She told it like this:

> In the fall of 1879, I purchased my first colony of bees of Rev. Willis J. King, who lived nearly a mile from our little home on the opposite side of the creek. They were the common little black bee, as that was the only kind in all this part of the country at that time, and were in a box-hive, as that was the only kind then in use hereabouts. My son (a lad of 13 years) and I went at night with a wheelbarrow after the bees. After securely wrapping them up with a sheet, we

started to bring them home. We had to cross [the Salado creek] on a swinging foot-bridge, but by me holding the hive and he rolling the wheelbarrow, we got them safely over. (Sherman 1896)

Footbridge over the Salado River

By 1886, she was writing for *The American Apiculturist* magazine, telling of her harvest of 2400 pounds of honey from 58 colonies of bees. "I would have taken double that amount if the season this year had been as good as it was last," she said. Mrs. Sherman also wrote extensively about beekeeping for women, placing her in the company of Jennie Atchley and Ellen Tupper, both of whom were successful writers and businesswomen. She pointedly wrote that over the years of her lifetime, women had gone from being relegated to being teachers or seamstresses, but occupations like beekeeping and small farming gave them a way to use their abilities and raise their families in the absence of husbands, many of whom had died in the Civil War. In her writing she also promoted beekeeping as a field rewarded by continued study and reading, but also as a path toward the greater appreciation of nature: "There are new beauties all the time coming to view. Even the despised weeds take on a new form of beauty never before dreamed of."

SOUTH TEXAS BEE-KEEPERS' CONVENTION, FLORESVILLE, TEXAS, AUGUST 10 & 11, 1900.

South Texas Beekeepers

About this time, she began purchasing Italian queen bees from the A. I. Root Company of Ohio, which had perfected the plan of sending them via U. S. mail. Mrs. Sherman was "the first person in this part of Texas" to do so. In her own words:

> I would recommend the Italian bees. They are far superior to the common native or black bee. I think that I am safe in saying that fully one-third of the native bees in this part of Texas have died this year from sheer starvation, while what few Italians that are here have yielded some little surplus with enough still in their hives to winter on while the blacks are still starving. A few such seasons as this and there would be no native bees left in this country. (Sherman 1896)

Of course, there were no native honey bees in Texas, but they were so common many people assumed they had been there "all along." Mrs. Sherman also attempted to propagate nectar plants but this proved to be more difficult. She tried large sunflowers, buckwheat, various clovers. She tried to get sweet clover on four occasions but it never took hold. She told of paying twenty dollars for "lucerne or California clover" seed (alfalfa). This grew about 8 inches and then succumbed to drought. Over the years, it has been deep rooted trees which are

191

the most dependable sources for producing honey.

In 1896, the *American Bee Journal* ran a long serialized autobiography filled with humorous and colorful anecdotes. She told a story of working with a hive, a gust of wind blew up her dress, and the ruffles infuriated the bees to the "extent I had never before seen and hope never to see again." She was able to put the hive back together but the enraged bees decided to go for two large, fat hogs nearby. By then the whole bee yard was raging, so she and two other women raised the pen so the hogs could escape the wrathful bees. It wasn't over yet, though. The hogs, rather than retreating, went straight to the bee yard. Sallie told how she and her father chased the hogs away from the bee yard, while the two other women moved the pen about 100 yards away, and they finally got the pigs to safety. The hogs were quite swollen from stings, but sustained no lasting injury.

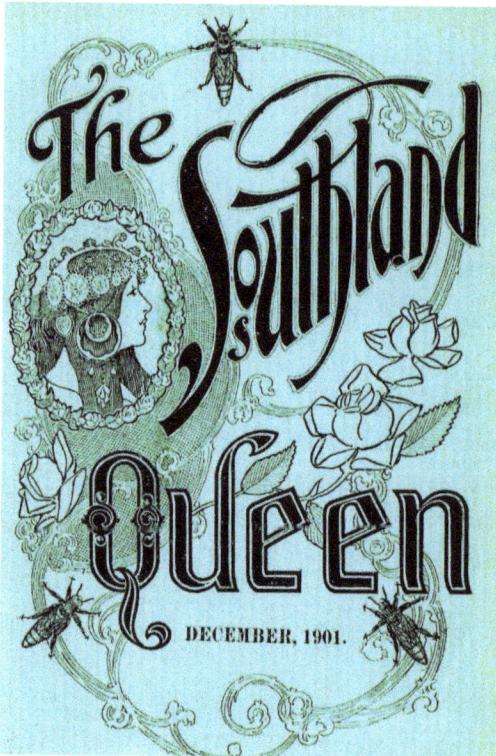

The Southland Queen magazine

At some point Sallie Sherman's parents required her full-time care, which naturally she gave to them. They had long since moved to Salado as well, both died in their 80s and were buried there. Not so much after, in 1901, Sallie succumbed at the young age of 58. It was her choice to be buried alongside her long-dead husband in a rural cemetery near where their farm had been. Years later, her son Charlie sent the following forlorn note:

My mother was a subscriber for the American Bee Journal for many years, and I think wrote for it occasionally. The Journal recalls many fond recollections of my youthful associations, and of my dear mother, whom I had the great misfortune to lose over 10 years ago. (DR.) C. H. SHERMAN. Dallas. Tex.

Works Cited

Anon. (1889). Mrs. S. E. Sherman. *Texas Farm and Ranch*. Vol. 8, No. 16.

Anon. (1902). Editorial Comments. *The American Bee Journal*. Vol. 42, No. 35.

Atchley, J. (1893). The Atchley Family. *Gleanings in Bee Culture*. Vol 21. No. 19.

Atchley, J. (1892). The American Bee Journal. Vol. 30, No. 10

Atchley, J. (1895-1904). *The Southland Queen*. Beeville, Tex.
Published by the Atchleys.

Bruckisch, W. (1861). Bee Culture. In: the Annual Report for 1860.
United States Commissioner of Patents.

Burleson, T. M. (1938). *Origin and development of the bee industry in Texas*.
Thesis, University of Texas.

Dumas, C. G. (1952). *Apiculture in early Texas*. Thesis, Southern Methodist
University.

Holley, M. A. (1833). *Texas. Observations, Historical, Geographical
And Descriptive*

Kerr, H. (1838). *A Poetical Description Of Texas*.

Sherman, S. E. (1886). Bee-Culture In Texas. *The American Apiculturist*.

Sherman, S. E. (1896). Fifteen Years' Experience in Bee-Keeping. *American Bee
Journal*. Vol. 36, No. 27-36

Sherman, C. H. (1911) The American Bee Journal. Vol. 60, No. 12.

Chapter 18

Pioneers of Texas Beekeeping

In the 1800s, Texas had so many industrious beekeepers, it is hard to choose where to focus. A list would have to include Judge W. H. Andrews, William R. Graham, W. R. Howard, E. G. LeStourgeon, Dr. W. K. Marshall, and T. P. Robinson at the very least. They won't all fit in an article of this scope, but I cover a few more.

Stachelhausen, the "Father of Texas Beekeeping"

Ludwig von Stachelhausen

There are some operations in beekeeping that are so common, that beekeepers never stop to wonder, how did this come about? I have in mind: the shaking of bees off their combs into a new hive, a procedure called "shook swarming." The purpose of the action is to prevent the natural swarming of the colony, which usually occurs at an inconvenient time for the beekeeper. This technique was championed by the well-known Texan, L. von Stachelhausen, though he was hardly the first to do it. Today, hundreds of thousands of hives have their bees shaken into boxes to be sold.

But to back up a bit, Mr. Stachelhausen was born in Regensberg, Bavaria in 1845, where he received a high level of education. He graduated from the University of Munich, Germany and continued at the School of Mines, in Loeben, Austria. By 1867, he was a superintendent of a glass factory back in Bavaria. About this time, he built up his first apiary and continued to keep bees for some 40 years. Stachelhausen married in 1870; the couple decided to move to the United States and settle in Texas.

Being well versed in science, progressive beekeeping, German, and English, he began writing extensively for the bee journals. Stachelhausen quickly developed a reputation as an innovator, although in the case of the "shook swarming," he was passing on a technique he had learned from the German beekeeper Gravenhorst, who in turn had copied it from the old fashioned skep (basket) hive beekeepers, who had learned to shake the bees from one hive to another as a means of propagating their bees.

An article in the *Beekeepers' Item* for November, 1922, stated:

> The County of Guadalupe, in Texas, has had a wonderful beekeeping history. It has produced many of the outstanding figures in the history and development of Texas Apiculture. L. L. Stachelhausen, known as the "Father of Texas Beekeeping" hailed from Cibolo in this county. (Garrison 1922)

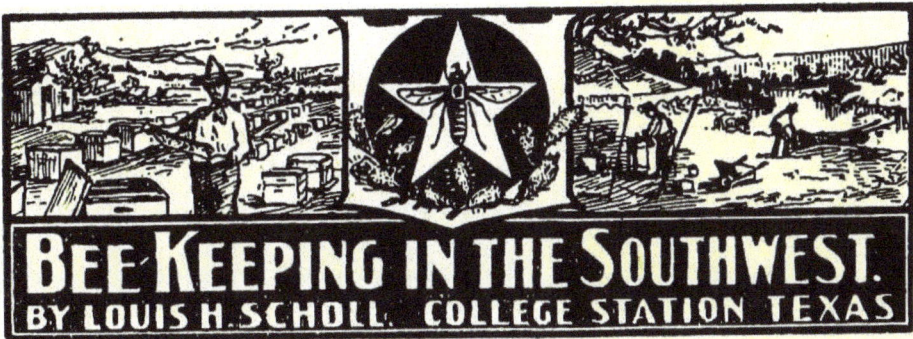

Beekeeping in the Southwest

Cibolo is not far from New Braunfels, home of a fellow Texas pioneer from Germany, Wilhelm Bruckisch. Another writer sent in this letter to the *American Apiculturist* magazine.

> The [article] on the "New Races and In-breeding" is worth three times the money of the yearly paper. All the contributions of Mr. Stachelhausen are an honor to your Apiculturist, for their plain, clear and correct language, free of all conjectures and only based on facts and experiments, and also free of all personal reflections as any article in a good paper ought to be. I found the writings of Mr. S. years ago in some German bee-papers, and his name has there a very good sound, and I was exceedingly pleased to find him here in America one of the ablest writers on bee-topics. (Kanzler 1890)

Stachelhausen gave a talk about "What are the Essential Qualities for Making a Successful Bee- Keeper?" at the Fourth Annual Meeting of the Texas Bee-Keepers' Association, at College Station, Texas, in July, 1904. He said, in part:

> *Baron Berlepsch, one of the most prominent bee-keepers of Germany in the 19th century (1815-1877) said: "At first learn theory or you will remain a bungler in practical bee-keeping all your life." His use of the word "theory" has another meaning than that for which it is generally used in the United States. Here [in the US] theory means merely a hypothesis not entirely proven. In Germany the word is used for science. To be a successful beekeeper the man must possess the necessary scientific knowledge, then and only then he will know always what to do if something unexpected happens in the apiary, and he will be able to improve his practical operations. (Stachelhausen 1904)*

Mr. Stachelhausen played a prominent role in the Texas Bee-keepers' Association. A friend of his wrote upon his passing in 1907, that they had lost one of the most highly esteemed members; at the meeting in July, at College Station, the members assembled half an hour in memorial service on Wednesday to commemorate his loss.

Texas A&M

> *What do you know of the past? Our good friend H. B. Parks, apiculturist of the Texas Experiment Stations, offers a valuable suggestion for the beekeepers of this State that ought to be carried out by all means and there is no reason why it will not be. The important subject is that of preserving those things of the past that have helped to make beekeeping history and which are and will be of immense interest. (The Beekeepers' Item 1920)*

There was interest in raising the educational standard throughout Texas. As early as 1839 Texas was interested in obtaining federal funding for a public university which would be available to everyone, not just the "well-born." Initially, agriculture was not included in the plans, but this changed. According to Henry C. Dethloff, in his *History of Texas A&M University*: (Dethloff 2000)

> *In 1853 the newly organized Texas State Agricultural Society included in its constitution a call for the establishment and endowment of an Agricultural College and model farm for the State, at or near the Capital, where the important business principles of Agriculture shall be scientifically taught and practically illustrated.*

The College was inaugurated in 1876. George S. Perry's *Story of Texas A and M* says the newly appointed director of agriculture, C. P. B. Martin, Doctor of Divinity, was "largely innocent of the scientific steps a farmer must take in the springtime," in order to have a crop to reap in fall. He lets him off the hook thoug,h because at the time there was no "appreciable body of scientific agricultural knowledge" in the United States. He goes on to say:

> The frank opinion of a majority of the people of Texas regarding this experiment in high-toned farming and mechanics was not merely that it was silly, but a piece of Yankee-instigated silliness. (Perry 1951)

Still, newly elected Texas Governor Richard Coke was determined to build an institution that would be affordable to students and at the same time give a good education in the fields of agriculture and "the mechanical arts." During the second year, the college proved so popular that it was overcrowded, with 331 students. Boys were sleeping in the hallways, some even in the president's office. Perry refers to A&M as a virtual outpost, "a place where wolves and deer roamed at night, where living conditions for the students were primitive." In the 1876-77 catalogs, the study of useful and injurious insects was specifically mentioned, but years would pass before formal entomology courses were offered. However, the fourth catalog announced lectures in bee culture. Van Allen Little wrote "A Brief History of Entomology at the A&M," from which I quote:

> The 27th Legislature appropriated money in 1901 for the establishment of an apiary at Texas A&M College for research and teaching purposes. Wilmon Newell was employed in February 1902 as an apiculturist in the department. He built a honey house and established an apiary. While he was here he wrote the bill for the first foulbrood law in 1903, and a course in beekeeping was offered for the first time. (Little 1960)

Louis Scholl

Louis H. Scholl was born in 1880 in Hunter, Texas; his parents were Germans. According to a brief biography in the *American Bee Journal* of February, 1910, his parents offered him no encouragement in his beekeeping pursuit and even took whatever money he made, as was the German custom. Further, they told him that beekeeping was not a real occupation and that he should learn a proper trade. Upon reaching 21 years, he left home and headed to Southwest Texas to work, eventually taking charge of some 1150 colonies in several counties. Scholl was made secretary-treasurer of the Texas Bee-Keepers Association and they had their first meeting in 1901 at College Station, TX .

Young Louis Scholl

In the summer of 1902, Scholl became a part time beekeeping assistant at Texas A&M. When Newell moved on, Scholl was appointed to the position as instructor of apiculture and was placed in charge of the college apiary. According to Little, "Scholl became known over the country as the man who wrote and talked about bees. Scholl's work and his enthusiasm did considerable for the development of this industry." In the Aug. 28, 1902 edition of the *American Bee Journal* (a weekly paper at the time) Scholl is quoted:

This is an apiary that not every State has, and we Texans are real proud of ours. I have had charge of it for some time, and we have fixed it up just like the good books say. Bee-keeping will now be taught the students who attend school here, and it is certain that some will go away with a good knowledge of apiculture, and will make successful bee-men when they launch out for themselves. (Scholl 1902)

MEETING OF TEXAS BEE-KEEPERS AT COLLEGE STATION, JULY 27-30.
Through the coöperation and perseverance of the members of the State Association the State legislature was induced to make an appropriation of $3000 for ridding the State of foul brood.

Meeting Of Texas Beekeepers 1909

In 1904, the *Journal* ran a lengthy report that Prof. Scholl presented at the 38th Annual Convention of the National Bee-Keepers Association, held in St. Louis, Missouri, with the title "Something About Texas Bee-keeping." He placed the number of Texas' hives at 400,000 and the honey crop at five million pounds. Scholl described the different regions and their flora, identifying Southwest Texas as "a country that I do not think can be surpassed anywhere." There one would find mesquite, "wahea" (guajilla), cat-claw, and a profusion of prickly pear cactus covering the prairie lands. This region was only fit for cattle and beekeeping, he said, and the coming railroads would enable its further development. Regarding his work at Texas A&M he described a four acre experimental plot for testing potential honey plants to ascertain whether they would be worth propagating in the region. Most of the Texas honey was coming from native shrubs and trees, although cotton plantations were a major nectar source.

As early as 1898, Louis Scholl was writing for Jennie Atchley's *Southland Queen* magazine, and he helped form *The Lone Star Apiarist,* in 1902, serving as its editor and treasurer. In 1905 he started writing for "Beekeeping in the Southwest" for *Gleanings in Bee Culture*, and the following year began a regular column in the *American Bee Journal* under the banner "Southern Beedom." In 1912, Professor Scholl published a 140 page boo,k *Texas Beekeeping*, which included everything from the equipment and skills needed, how to set up a honey house, the nectar flora of Texas, and even a section on binding bee journals into yearly books. He started his own bee journal, *The Beekeeper's Item*, in December, 1916, at New Braunfels, Texas and was its editor and publisher until 1926. During that time, his wife had a regular column in the magazine, assisted by their daughter Margarethe. In it, Mrs. Scholl relates how the family hit upon the idea of setting the eight year old girl up in a small business. She was to be paid two cents for every mouse she trapped about the property including the honey house, the garage, etc. Not only was she an efficient trapper, but she kept an impeccable account book.

An obituary reads, in part:

> Louis H. Scholl served in the Texas House of Representatives from 1915-1917, representing the Comal and Hays Counties area of Central Texas. The Volunteer Firemen of Texas sponsored a plaque on Rep. Scholl's grave extolling his service to the State Firemen & Fire Marshals' Association of Texas, as well as his work as a legislator, educator, and more. They commended Rep. Scholl as "the state's authority on bee culture and a tireless civic leader." Representative Scholl passed away in 1956.

H. B. Parks

H. B. Parks, Apiarist

Although he arrived in Texas too late to be involved with the beginnings of beekeeping at Texas A&M, H. B. Parks is an important figure in its story. Harris Braley Parks was born on June 10, 1879 at Carlinville, Illinois. He received his B.S. degree there at Blackburn University in 1900 and worked as a teacher at Sitka, Alaska, from 1907 to 1911. Parks moved to Texas in 1917 and by 1920 he was apiculturist at the Texas Agricultural Experiment Station. H. B. Parks was a prolific writer for the magazines such as *The Beekeepers' Item* and the *American Bee Journal.*

His article in the May, 1917 issue of the latter was titled "Why Some Beekeepers Fail."

> *To sum up the causes of failure it was because the would-be beekeeper did not know of the whys and hows of the bee trade. One did not know how to buy supplies, another how to handle the swarms, another how to sell honey, another where to place the hive, and so forth. All were not inclined to study the subject, for everybody has sense enough to keep bees. All those that have succeeded have studied the cause of former failure and keep in tune with the bees, and other bee-men, are enjoying their labor and the sweets thereof. (Parks 1917)*

In 1919 he published a brochure *Beekeeping for Beginners,* tailored to the needs of new beekeepers in Texas. A short biography of Parks appeared in *The Beekeepers Item*:

> *To the readers of beekeeping literature, agricultural papers and a number of other classes of publications, the name of H. B. Parks is a most familiar one.*
>
> *He is always a prominent figure in beekeepers' gatherings. Although his serious outward appearance does not reveal the humorous inner man, this comes to the surface in conventions. The mixture of serious [and] humor has much to do with making him a most likable member of the fraternity, and it is therefore no wonder that he is so*

well known as "Honey Boy" Parks among his more intimate friends.
(LeStourgeon 1923)

In another issue, Mr. Parks was featured in the recurring column "Some
Beekeepers of Texas," penned by Guy LeStourgeon, another prominent writer.
Next to a photograph of H. B. he wrote:

Look at this picture and note the evidence of wind. There is one
characteristic of H. B. Parks that has become an axiom in Texas
beekeeping circles. That is his breeziness. I was once present when
a stranger met Mrs. Parks. He said, "So you are Mrs. Parks. Tell
me, do you ever get a chance to say anything? I have heard that
if your husband would start in now and try to repeat everything
he has said that he would beat old Methuselah's record for years
before he caught up with himself." And look what talking–just plain
word spilling–has done for him. He has chatted and lectured and
harangued himself into more jobs than a dozen men ought to have.
He works off his surplus energy by writing special articles for a half
hundred beekeeping, agricultural and technical papers. And then, in
intervals of rest he just simply talks. (LeStourgeon 1921)

Apiary at Texas A_M 1902

In 1922, Mr. Parks established an Apicultural Laboratory near San Antonio.
By 1945 he was semi-retired and moved back to College Station, where he

became curator and botanist of the Tracy Herbarium at Texas A&M. Parks' chief interest was bees, which he collected extensively, especially in Bexar County. In collaboration with V. L. Cory, he published *Flora and Fauna of the Big Thicket Area* (1936), the first comprehensive survey of that area. He also worked with Cory to compile the first catalog of the vascular plants of Texas, *Catalogue of the Flora of Texas* (1936). Parks also wrote and published *Valuable Plants Native to Texas* (1937). Harris Braley Parks died on November 19, 1958, at San Antonio and is buried at San Jose Burial Park San Antonio, Bexar County, Texas.

HONEY MAP OF TEXAS,

Honey map of Texas

There was another side to H. B. Parks: he collected stories. In the publication *Texas Folk And Folklore*, he gave the first account of the origin of the Black American folk song "Follow the Drinking Gourd." Parks told of his travels from living among native people in Alaska, to time spent in Hot Springs, NC, the Big Rich Mountains on the border between North Carolina and Tennessee, eventually wending his way to Texas. As interesting as that story is, there is another, published in 1930, about the legends surrounding the Texas Honey Caves, where supposedly enormous honey combs are guarded by hordes of rattlesnakes.

Works Cited

Anon. (1902). Editorial Comments. *The American Bee Journal*. Vol. 42, No. 35.

Anon. (1907) Death of Stachelhausen. *Gleanings in Bee Culture*. Vol 35, No. 18.

Atchley, J. (1893). The Atchley Family. *Gleanings in Bee Culture*. Vol 21. No. 19.

Atchley, J. (1895-1904). *The Southland Queen*. Beeville, Tex.
Published by the Atchleys.

Burleson, T. M. (1938). *Origin and development of the bee industry in Texas*.
Thesis, University of Texas.

Dean, W. H. (1912). Texas - Land of Honey. *The Texas Magazine*. Houston, Tx.

Dethloff, H. C. (2000). *A Centennial History of Texas A&M University*, 1876-1976.
Texas A&M University Press.

Dumas, C. G. (1952). *Apiculture in early Texas*. Thesis, Southern Methodist University.

Garrison, C. W. (1922). The Beekeepers' Item. New Braunfels. Texas. November.

Jeffers, J. (1936). Caves of Honey. *Gleanings in Bee Culture*.
Vol 64. No. 11.

Kanzler, W. F. (1890). The American apiculturist. Wenham, Mass. February

LeStourgeon, E. G. (1921) "Some Beekeepers of Texas." The Beekeepers' Item, January.

LeStourgeon, E. G. (1923) "Beekeepers Here and There." The Beekeepers' Item, February.

Little, V. A. (1960). *A brief history of Entomology at the Agricultural and Mechanical College of Texas.* Publ. College Archives, Texas A&M College.

Parks, H. B. (1917) "Why some beekeepers fail." American Bee Journal. Vol. 57, No. 5

Parks, H. B. (1928). Follow the drinking gourd. Publications of the Texas Folk-Lore Society, 7, 81- 84.

Perry, G. S. (1951). *The Story of Texas A and M.* McGraw-Hill. New York. Scholl, L. H. (*editor* 1919-24). *The Beekeepers' Item*, New Braunfels, Texas.

Parks, H. B. (1917) "Why some beekeepers fail." American Bee Journal. Vol. 57, No. 5

Parks, H. B. (1917) "Why some beekeepers fail." American Bee Journal. Vol. 57, No. 5

Scholl, L.H. (1902). The American Bee Journal. Vol 42 Issue 35

Stachelhausen, L. L. (1904). Fourth Annual Meeting of the Texas Bee-Keepers' Association, at College Station, Texas, in July.

Weaver, B. A. (1982). History of the Zach Weaver Family and Weaver Apiaries. In: *The History of Grimes County: Land of Heritage and Progress,* ed. ed. Grimes County Historical Commission. Dallas, TX.

Chapter 19

The Legendary Honey Caves of Texas

In the previous chapter, I mentioned that H. B. Parks was an avid collector of tall tales and folklore. Parks compiled the honey cave legends in an article published in 1930, and that collection is a great read. However, I will tell this story in my own way. Some of what follows may be true and some not, but I leave it to you to discern.

Campfire Talk

Sitting around the crackling campfire, trying to stay upstream of the thick smoke of mesquite wood, the talk ranged from armadillos to road runners. They called that bird "paisano" and more than one person would swear to having seen them build a cage out of cactus branches to trap a rattlesnake. Snakes were always a peril, anything that could kill a man with one bite was not to be trifled with. Back in those days, hunting was a favored pastime and some found special delight in raiding wild honey bee nests. Anything sweet was precious then, so men thought nothing about felling tall oaks to wrest the combs away from the angry bees. And also caves.

It seems like everybody had a story about a honey cave they had glimpsed or heard of, or knew somebody who knew where. A group of four young men told of going after wild bees, up the Lipan Creek canyon. They described an enormous bee-hive in a south facing cliff, from which they and three other families got a supply of honey for years. The entrance is some forty feet above the creek-bed, where there is a long crack in the rock hundreds of feet long but only wide enough to crawl into. Back in there is hive upon hive, the nests and the honey they have been hoarding for so many years. So much honey is in there that on very hot days, the honey weeps and oozes out of the cracks and seams of the sandstone. You can taste honey in the creek water a mile or two downstream. The bees spend a good portion of their time retrieving the dripping honey and carrying it back to the huge hives.

Summer is not the time to go bee-hunting for the air would be almost black with clouds of bees which can be seen and heard miles away. The honey hunters preferred to wait for cold weather, near Christmas, when the bees are groggy from the chill. The particular time of the story was especially cold, being that there was a stormy "norther" approaching and they took shelter in the lee side of the cliff, near where the bees were. Nobody tries to do anything when a norther comes on, you can't even get a blacksmith to shoe a horse till it blows over, often after two or three days.

Texas Cave with tourists

So they huddled down and built a big fire to wait it out. After a while the fire had died down and they were waked by falling rocks and an infernal growling. Soon they were engaged in hand to hand combat with a huge bear. Anse grabbed an axe and bashed the growler's head, Charles got to his feet and unloaded his gun into the bear, killing it dead. His buddy Wert shouted: "Be careful where you shoots! Whole dozen dem buck shots go buzz past my ear!" Pretty soon another bear stuck his head out of the cave entrance and began climbing down the cliff. Every minute or two a head would pop out from the crevice.

When daylight came they gathered logs and built a scaffold to reach the cave. They managed to reach it and got sight of a huge black cavern, from which low growling could be heard. No luck shooting at the bears in the dark, so the hunters gathered more wood and threw it into the cavern, started a bonfire which lit up the whole bear den. The bears were cornered in their lair. To shoot them like this might seem unsportsmanlike hunting, but what the heck. They shot and hauled out seven dead bears, to add to the others already laid outside. Fat bears, too, with a noticeably sweet taste, having been plumped up on so much honey.

Bee Mountains

Bee Rock, Clifton Texas

Mr. J. W. Hornbeak was born in a log cabin about two or three miles from "the mountain" which is located on the Bosque River, not far from Meridian. Back in the day, the Indians would come down from the north after a wet spring to get honey from bees nesting under ledges at the top of the cliff, some seventy feet above the river. By cutting and splicing mesquite poles, some of the lighter folks would shinny up, gather the honey comb and lower it down by a rope attached to a grass sack.

When Hornbeak was a boy about ten, a group of hostile Indians came for honey and after they had gotten as much as they wanted, they worried the white settlers might catch on, so they killed as many of them as they could. Hornbeak and them what survived moved off to Navarro County. Years later, visitors to the site would still find hundreds of mesquite poles which had been used to construct ladders.

Another legendary bee cave is supposed to be close by the city of San Marcos. There is an immense bluff full of holes in which reside, not only giant hives of bees, but hundreds of swarming, writhing rattlesnakes. Heard tell about 1885, an adventurer from the East rounded up a group of men from Austin to rip open the side of the bluff. The Easterner had been told that many people from San Marcos had squeezed inside in the summer when rattlesnakes weren't at home. They told of hundreds of pounds of honey and wax, and sporting a patented smoke gun, he aimed to get it. Upon descending to a depth of thousands of feet, he emerged with tales of massive sheets of honey combs hanging from the cavern ceiling.

The men from San Marcos set to enlarging the entrance to the cave with sledge hammers and chisels. But no sooner had they started banging on the rocks, then snakes began to pour from every hole and crevice in the side of the cliff. Although none were bit, the sight of so many snakes struck fear in the men. They beat a retreat, and no promise of hidden treasure would be enough to get them back.

In the Davis Mountains, is said to be another cave which opens through a vast entrance like a huge cathedral. The bees come and go in clouds, from this great opening. One can easily see for oneself the curtain like sheets of honeycomb hanging down from the ceilings. As far back as anyone has gone, there are strikingly white combs, full of honey, many feet in length. But beware, this cave is guarded by supernatural forces. Once an adventurer is deep inside the cave, he begins to feel his feet entangled in bones, the remains of so many human skeletons. Should he fail to take heed, and continue further into the cave, he is crushed to death by unknown forces. Anyone who follows to retrieve him suffers the same fate, and succumbs to the dark force which protects the Mountains' honey bees.

The Devil's Punch Bowl

Why, I even heard of a huge cave in the brakes of the Devil's River which has enough honey to make any man rich. The cave is called "The Devil's Punch Bowl," and according to locals the bees swarm out so thick that they look like a great smoke signal from two or three miles away. This part of Texas is so warm that there are flowers pretty nearly year round and when there isn't they feed off of berries and prickly pear fruits.

And why hasn't all this honey been taken yet? As it happens, it's rough country, and inaccessible by anything but hiking and packing provisions on the back of a burro. The sinkhole itself is hundreds of feet deep and honey hunters have been stung near to death hanging on a rope halfway from the bottom. Perhaps the biggest problem is getting the honey out of the back country. It is usual to strain it and put it in honey cans which hold about two gallons each and are paired to accommodate the size and shape of a burro. But get this, once the animal is loaded down with the bounty, the prospector is now tasked with carrying his gear on his own back. There's a long distance between water holes and decent camping spots, the whole trip of 135 miles can take two or three weeks.

Queen's Canopy

Nevertheless, a group of hunters from Kansas City on the prowl for deer and bears, heard about the Devil's Punch Bowl and decided to have at it. The Wisconsin Naturalist Journal describes what happen next:

> When they reached it, like everyone else seeing it for the first time, they were amazed at the proportions of the wonder; a hole forty feet in diameter yawning open in the middle of a wide valley, with a perfect torrent of bees rushing up from it like dirt blown from some mighty blast and all the while a roaring as loud as a great cataract; looking down into the great abyss, for the hole widens immediately below the surface, they saw the festoons of honey hanging there which the bees had strung along the sides of their mammoth hive after they had filled the hidden grottoes; and looked in through the upward swarms and saw the gleam of combs built no doubt many years before. (Anon 1900)

As you may imagine, the Kansas City gentlemen decided to form a company to mine the honey and get it to market. They appointed a Mr. Ouden to organize the whole project. He went down to the Punch Bowl but came back with the report that they had best hire someone else, so their idea was abandoned. Meanwhile, Ouden went back down, bought the land, and a big derrick, rounded up about 40 men and 100 pack mules. His plan was to send men down wrapped in mosquito netting, suspended from the derrick, in order to fill boxes of honey and pull them out with a pulley. To make a long story short, Ouden ended up in the bottom of the sinkhole and the men absconded with the equipment. How he managed to get out of there has never been learned but it was said that the very mention of honey would make him sicken and run.

Notes on the Legends

As I said, there is some fiction and some truth in these stories. For example, the first story appeared in 1881, in a magazine called *The Youth's Companion*. There is a nice illustration of the young men and the scaffold they built to reach the cave. Mr. H. B. Parks related the story of Mr. Hornbeak of "Bee Mountain." Parks visited the location in the 1920s, and assured his readers that while there were caves of bees, the season had been so dry that none of them would have been expected to contain much honey. I have included photographs of the imposing cliff.

But beyond that, by the time Mr. Parks began visiting the many Texas caverns, it had been learned that some of them were quite vast and bearing immense limestone formations which perhaps in dim light would appear quite similar to great sheets of white honey comb. But these are caused by decades of seepage, which formed the eerie and beautiful shapes seen in the caverns, and are a wonder to behold even if not made of wax and honey. As Parks points out, the damp and seeping water would make the caves unsuitable for bees, even if they chose to live in them. He claimed to have seen piles of damp moldy comb in some of the dark hollows.

My research found many interesting references to the caves. A very early printed reference states: "In the adjoining counties among the *mountains*, are bee caves. Some of them contain tons of honey when found. MRS. J. L. CUNNINGHAM. Strickling, Texas, June 19th, 1877." And this: "I see you speak of bees in caves; they do work in caves, for I have robbed them of barrels of honey at a time. That was on the Devil's River, Texas. I did not know anything about their work, whether they had any queens or not, but I do know that they had lots of honey in their caves. H. R. C. BREECE. Greenwood, Col., March 23, 1879."

Jennie Atchley, editor of The *Southland Queen*, attempted to dispel these tall tales. She wrote in 1892 for the *American Bee Journal*:

> I suppose you have all heard of the bee-caves in Texas, where the bees work through an orifice in the rock, in a stream as large as a flour barrel, and where wagon-loads of honey have been taken, etc. These statements somehow or another get magnified terribly by the time they reach the press. There are strong colonies of bees that occupy caves, I will admit, but there are no more bees there than in any other strong colony, as there is a limit to their strength. As we all know, there is only one queen, or perchance two, as in some instances in our hives, but one of them is usually old, and of no value. The bees in these cliffs are usually hived in a large crevice or crack in the bluff – very often not more than a foot deep, and at other times the combs are built clear out on the outside of the rock. Bee-Caves in Texas – Mistaken Ideas. (Atchley 1892)

But that stopped no one. In 1900 *The Wisconsin Naturalist* ran with this:

Caves of Wild Honey. Sweet Store House In Texas. Bees Work the Year Round and Have Gathered Tons of Their Product Far Away from Civilization.

There is enough honey in the brakes of Devil's River (Texas) to make any man rich who can get it to market. There are tons of it; in clefts in the rocks, in hollow trees, in caves and in the famous "Devil's Punch Bowl," which is a great sink in the Devil's valley and out of which bees always swarm in clouds so thick that at a distance of two or three miles it has the appearance of a great signal smoke. (Anon 1900)

Bats leaving the Devil's Sinkhole

A visit to the "Devil's Sink Hole" today will expose the truth of the huge clouds of bees blackening the sky and seen from miles away. These caves are filled with bats, not bees. According to the State Park, "Some 3 million bats emerge in a swirling mass from the Devil's Sinkhole in search of food on warm nights. Marvel at the amazing spectacle at this state natural area northwest of San Antonio." While the sink hole is basically a great big hole in the ground, other Texas caves are most various and beautiful. You won't find bees and honey there, but neither rattlesnakes nor bears. Some of the lesser known and more inaccessible caves have been found to contain skeletons. In the beekeeping magazine *Gleanings* of 1920, Mr. Parks himself sought to alert gullible readers:

The annual bee-cave story is again in print. This time the cave is located in Menard County, Tex., and contains acres of solid comb honey. The bees in a solid cloud and with a roar like that of distant thunder leave and enter the cave. These bees collect this store of honey from the Rio Grande Valley and Mexico (only 150 miles away). A company is being formed to drill wells into this cave and pump out the honey. The story came to us from a Seattle paper, and shortly afterward a lawyer in Ohio wrote us that a client of his wanted information about the cave as he was about to buy stock in it. Let us warn the public that; while there are numerous small bee-caves in the limestone hills of Texas, the above story is a hoax and any such company is unknown here. If you must buy stock in wildcat schemes, try oil and you will not get stung – at least by bees. (Parks 1930)

The awe-inspiring surfaceward view from the top of the great rock mountain of the Devil's Sinkhole. Photo by Bill Helmer. Courtesy of Mills Tandy.

Bottom of Devil's Sinkhole

I wish to conclude with this excerpt from "Some Legends of Texas Pioneers" by Stella Gipson Polk, written in 1974:

> *So the legends ran; yet it was no legend that many a man extracted the purest of honey made from whatever bloom was prevalent in his section of Texas: chaparral, bee brush, clover or guajillo. (Polk 1972)*

Works Cited

Anon. (1881). A West Texas Adventure. *The Youth's Companion.* 44(41).

Anon. (1900). Caves of Wild Honey. *The Wisconsin Naturalist.* 6(2).

Atchley, J. (1892). Bee-Caves in Texas – Mistaken Ideas. *The American Bee Journal.* 30(11).

Parks, H. B. (1920). In Texas. *Gleanings in Bee Culture.* 48(12).

Parks, H. B. (1930). The Lost Honey Mines of Texas. *Southwest Review.* 16(1).

Polk, S. G. (1974). Some Legends of Texas Pioneers. *The Cattleman.* 61(7).

Texas State Parks. Devil's Sink Hole State Natural Area.
https://tpwd.texas.gov/state- parks/devils-sinkhole

Chapter 20

Pioneers of Texas Beekeeping

To conclude this series on the early history of beekeeping in Texas, I can't overlook the Weaver family. They have long been known worldwide as producers of queens and packaged bees for sale. Their beginnings in bee culture go back to the time of the other Texas pioneers.

The Pioneering Weavers

Weavers family 1897

Florence Somerford was born in Florida, November 14, 1867 and moved to Texas with her family in 1872, settling near Navasota. She married Zachariah S.

Weaver in 1888, and they rode horseback to their new home where she lived for more than 70 years. As a wedding gift, her brother Walter Somerford gave the couple ten hives with bees. Florence was already a seasoned beekeeper, having worked with her brother. The following appeared in the magazine *Gleanings in Bee Culture*:

> *My eldest brother, Walter, who is 18 years old, has 10 hives of bees. He has two smokers, one Clark's cold blast and the other a Quinby. He has his bees in two apiaries, one at home and the other at Linn Grove, four miles off, where there are plenty of linn-trees. The bees are building queen-cells, and fixing to swarm. They are gathering lots of pollen from cottonwood, willow, dogwood, and sassafras. Bees commenced gathering pollen the 2d of January, from elm. There are not many people who keep bees around here. Rosalie E. Somerford. Navasota, Texas, March 11, 1887. (Gleanings)*

1898 Weavers Lynn Grove apiary

We may assume Florence's sister Rosalie is referring to linden trees, or basswood. In 1891, Walter himself wrote to the magazine: "Basswood Honey in Texas. You ask where we get linden honey so far south. Why, friend there is a section of linn hummock that is as fine, I suppose, as any you ever saw. Some trees are 3½ feet in diameter." The Somerfords and the Weavers became extensive beekeepers in the region. In May, 1912, the *Texas Magazine* had this to say:

> *For twenty-three years Mr. Z. S. Weaver has been a beekeeper. His home and apiaries are located at Courtney, Texas. During these years he has had two failures in the honey crop. For ten years his average production of surplus honey was 100 pounds extracted to the colony; for eleven years 75 pounds per colony. Mr. Weaver conducts a general merchandise business and says: "I find that bees pay a better dividend than any investment I have ever made." Mr. Weaver takes the best of care of his bees, and in return they do their very best for him. (Dean 1912)*

In 1892, quoted in the magazine *Gleanings in Bee Culture*, Zach quipped: "I have no short cuts, but find that that I get well paid for all fussing that I can do." This next excerpt is from J. E. Davis, of Navasota, writing in the *American Bee Journal* in 1961:

A Texas Story

> *One spring we were working in the honey house at Navasota, Texas, putting in foundation, when Grandma Weaver came in, pulled up a hive body, and started talking:*

> *"Well, I'll tell you all about the time my husband and I went over to our sunset yard to extract honey. In those days we did everything in the bee yard. We got there a little before noon so I fixed a small lunch while Mr. Weaver was staking out the oxen and setting up the extractor and barrel.*

> *"After lunch we went to work. Late in the afternoon, one of those Texas spouts came up and it rained for about two hours. Extracting was over. The oxen were hooked back up and we headed home. We got back to Brushy Creek, it was out everywhere so we just sat there and let it run down. As soon as it was low enough Mr. Weaver started the oxen on and told me to steady the barrel and the 10-gallon can but during the rundown of the water a log had been washed across the road and the oxen were stumbling and the wagon almost turned over so I lost the 10-gallon can. I guess the fish had their fill of honey for supper." (Davis 1961)*

A photograph of the Weaver family taken in 1897 features Florence and Zach with five of their children. In the right front is Roy, who was born in 1892 in Lynn Grove, and married Lela Binford there in 1916. Their son Binford credited Roy with turning the bees into a commercial enterprise, which he continued. About his grandmother, Binford Weaver wrote:

> *Optimistic, and with a great zest for life petite Florence was a true pioneer, willing and also able to do what had to be done for her own family or someone else's family. She raised or partly raised 32 children, only eight of them her own. And when she died at the age of 101, her wedding dowry of 10 colonies of bees had become legion. (Weaver 1982)*

Roy Weaver

Dne of the families of Texas famous in beekeeping is the Weaver family of Nav- isota. Its history dates back to 1888. (Left) Roy S. Weaver, Sr., his three sons, Roy S. Weaver, Jr., Dr. Nevin Weaver of College Station, and Binford Weaver; Ioward Weaver (brother to Roy, Sr.) and his son, Billie Howard.

Weaver family 1958

According to an article published in 1958, a big change came in the mid 1920s for the Weaver Apiaries. Roy began managing the bees in 1915 and caught on to migratory beekeeping. The cotton fields of the Texas blacklands was the place to make big crops of honey so they took to moving their hives there. Things took an abrupt turn when growers started to use arsenic heavily to control pests. In Roy's words:

> If it had not been for a word of encouragement with some practical assistance from my friend, Horace Graham of Cameron, we probably could not have survived in the beekeeping business. Graham had been urging me to go into the queen-rearing business, and when he offered to buy a minimum of 1,000 queens a year from me, I began operations. That was in 1926. (Weaver 1982)

Pure Italian Queens and Package Bees for 1927

Why lose those queenless colonies? Why keep those old, worthless Queens?

Book your order now, have your Queens and Bees sent when you want them. Two-pound package, $2.50; Queens, $1.00. Special prices on large quantities.

Roy S. Weaver, Navasota, Texas

Roy and his brother Howard set up a partnership, raising Italian and Causasian queen bees. The home base was not very suitable for honey production but proved ideal for queen rearing. The South Texas climate is mild in early spring and a series of minor plants provided a nectar flow of sufficient duration to build up the colonies to be able to produce queens and bulk bees for sale. Roy's words again, from 1958:

> Trucks carry from 500 to 1,000 packages, although one big load drove off with 1,120 packages on board. At one time, all package bees went by express, but these days, about 75 per cent go by truck, 20 per cent by express, and five per cent by mail. (Weaver 1982)

Nevin and Elizabeth Weaver

A less well known member of the family was James Nevin Weaver. Roy Weaver had three sons: Binford, Roy, Jr., and Nevin. The first two went into the bee business but Nevin's path was different. While doing his post-doctoral studies at Harvard, he met and married Elizabeth Chadwick, who was born in Cambridge, Mass. They moved their family to College Station, Texas, where Nevin became a professor of biology at Texas A&M University.

In 1952, A&M was one of the first universities in the nation to receive research funding from the National Science Foundation, some of which went to Prof. Weaver's research into honey bee nutrition. His work included developing techniques for rearing honey bee larvae on royal jelly in the laboratory, and studies of the effects of larval age on dimorphic differentiation of the female honey bee (the transformation of the female bee egg into either a queen or a worker depends not only on nutrition but the timing of the feeding).

Nevin and Elizabeth co-authored contributions to *Bee World,* a magazine published in England since 1919. They spent a great deal of time in Mexico between 1958 and 1978 and wrote a lengthy description of beekeeping with the stingless bee *Melipona beecheii,* called *colecab* by the Yucatecan Maya. Here is a brief excerpt:

> *The actual work with bees cannot be separated from the rituals that accompany it. Before he touches a hive, the beekeeper scrubs his hands and arms thoroughly with water and the leaves of chacah (Elaphrium simaruba). We seldom learned much about the special place of colecab by direct questions, but several observations indicated that the beekeepers will go to great lengths to avoid killing bees. If a bee is accidentally killed, it is folded in a bit of leaf and buried. (Weaver 1981)*

Elizabeth Weaver continued her interests in science by serving as a research assistant with a biology grant in Yashcaba, Mexico; together they hand built a house on the family farm in Navasota, Texas. In 1965, the Weavers moved to Massachusetts when Nevin accepted a faculty appointment at the University of Massachusetts at Boston. Elizabeth was active in liberal causes, photographed courthouses and wildflowers all across Texas, and hiked and camped in the Big Bend region of western Texas. Nevin retired in 1990 and passed away in 1995 in Cape Cod, where they spent the summers. Elizabeth lived to be 87, and died in 2008.

Works Cited

Davis, J. E. (1961). "A Texas Story." American Bee Journal. Vol. 101, No. 5.

Dean, W. H. (1912). "Texas–Land of Honey." The Texas Magazine. Vol. 6, No. 1.

Moffett, J. O. (1979. Some beekeepers and associates. Stillwater, OK.

Weaver, B. (1982). "A History of the Zach Weaver Family and Weaver Apiaries." In: *The History of Grimes County: Land of Heritage and Progress.* Grimes County Historical Commission.

Weaver, N., & Weaver, E. C. (1981). "Beekeeping with the Stingless Bee Meupona Beecheii, by the Yucatecan Maya." Bee World. Vol. 62, No. 1.

Chapter 21

The Turn of the 20th Century

People look backwards to history to give perspective on where we are now. How far have we travelled? Did we fulfill our goals and dreams? Most readers will remember "Y2K" and its accompanying fuss and bother. What about the moment that Americans left the Nineteenth Century and entered a brave new era of modern living? In this chapter I intend to focus on the changes that took place in the beekeeping world at the end of the 1800s and the beginning of the 1900s. In previous chapters I have covered such topics as the transition from horse and buggy to the automobile and the tremendous impact this had on beekeeping, for better or worse (by this I mean, increased production brought with it plummeting honey prices).

The 20th Century

> *Your committee could not have struck a subject more difficult to write upon, and yet of such live interest to the progressive beekeeper of the new century. Mr. Hutchinson calls it the most hopeful field to which beekeepers can turn their attention, and says "keep more bees" which means "out apiaries." There was a time when extracted honey brought from 15 to 20 cents per lb., and comb honey was proportionately high; when the forests were plentiful and filled with basswood (that grand honey tree), now almost extinct. These forests, in addition, contained and retained moisture, and drouths were less frequent in consequence. (Sibbald 1901)*

Except for the prices this might sound familiar, but it was written in 1901. The writer of yore bravely lurches forth:

> *This seems to be an age of specialists, and our thought, time and attention must be concentrated upon one thing to make the greatest success of it. This can only be done, and the above noted changed conditions met by increasing the number of colonies kept, so as to make provision in the fat years against possible lean ones. But the difficult part is to tell you how to do it, and a saying I've heard comes*

to me and seems quite appropriate. It is "The more I know, I know I know the less," and so venture to give help on this "new hopeful" field with fear and trembling. (Sibbald 1901)

Extracting the honey

SESPE APIARY, OWNED BY J. F. McINTYRE, AT FILLMORE, CALIF.—LOOKING WESTWARD.

Sespe Apiary, Owned by J. F. McIntyre, at Fillmore, Calif. Looking Westward

The modern frame hive is essentially a product of the 1800s. The original frames were ¾" wide and had to be spaced by hand. Julius Hoffman introduced frames with wide end bars which kept the frames at the correct distance apart. Another improvement was the use of wider and thicker top bars, which were less prone to sag or bow. When his newfangled frames came out, they were criticized of course. Hoffman replied:

> Mr. J. A. Green, Oct. 1, 1904, says that the Hoffman frame is not the frame for average bee-keepers, as they will forget to crowd the frames together when finishing work in a hive. This is easily answered by giving such average bee-keepers the advice to go back to the old box hive. However, I for one have a better opinion of the average bee-keeper. I know many of them personally who are not slovenly, but handle the Hoffman frame all right. (Hoffman 1905)

Customers wanted honey in the comb, so a variety of boxes were invented to cater to this market, but beekeepers realized they could produce far more honey if it could be extracted so that the combs could be reused. The hand cranked extractor was the tool of choice for many decades. Combs would be uncapped with a broad knife and loaded into the machine to be spun. The process was quick and it seemed there were always youngsters who could be induced to turn the crank for hours, days and weeks.

The custom in the late 1800s, which continued well into the 20th century, was to extract in or near the bee yard. Often the emptied frames were immediately returned to the hives. When beekeepers began to keep bee yards miles distant from home the practice was either to haul the honey extractor to the bee yards or to keep one extractor in each location along with a shed in which to extract and to store equipment so that practically everything needed would be there. The crew could arrive on bicycle, get the work done, and bike home. Often the honey would be stored in barrels in the shed until after snowfall, when they could be hauled home by means of a horse drawn sleigh.

Some large beekeeping operations had hundreds of hives in one location, especially in California. Possibly the earliest recorded instance of a motorized honey extractor was an article about J. F. McIntyre, in the *Pacific Rural Express* newspaper, Oct. 1891. The apiary, located in Ventura County, was situated on a gradually sloping hill. McIntyre used gravity in a number of ways. First, the supers were carted down to the honey extracting building. Water was available from a sluice that descended from the mountains into the valley to irrigate oranges and lemons. A water pipe ran to the extracting facility where it was connected to a "Pelton Wheel" which is basically a water pump in reverse, where the water pressure turns the turbine and provides power to drive the honey extractor. Finally, gravity would move the honey downhill in pipes to large holding tanks.

FIG. 1215. NO. 0 PELTON WATER MOTOR.

Pelton Water Motor

The Pelton water wheel was invented by Lester Pelton. He had been lured to California by the Gold Rush. In the 1860s steam engines were used for mining operations but these consumed great quantities of wood. Pelton saw that water power supplied from small local reservoirs could be harnessed to drive machinery. He collaborated with Miners Foundry of Nevada City, CA, which is still in existence as the Miners Foundry Cultural Center (visit minersfoundry.org).

'British 'Weed' Foundation Factory

British Weed Foundation Factory

Large operations with plentiful fuel resources used large steam engines for power, but a real breakthrough was the small gasoline engine. Early prototypes were in use by the 1800s with a flurry of patents issued in the 1870s. These were much better small motors which could be run and maintained on the farm for various purposes, powering pumps, mills, saws, and so forth. In the cities, electricity was being supplied to households, businesses and factories. This was generated by steam turbines, gasoline engines, as well as immense hydro-electric power plants driven by water sources such as Niagara Falls. Electricity was powering all sorts of equipment in the cities but it took much longer to reach rural America, so many people used small gasoline engines to generate electricity right there on the farm.

This was a time of great innovation but the hazards of mechanization were plenty. It was common for workshops to employ single power sources, whether water wheels, large steam engines, or the like. The power would be distributed via overhead shafts in the workshops and factories. Motion was transferred from the shafts with belts and pulleys, so a typical workspace would be treacherous due to constantly moving wheels, belt drives, saws, grinders, and so on. Safety guards and shields were rarely in place where they would be today.

Uncapping the Combs

Left: Interior of the Truck Body, Showing the Motor-Driven Extractor Used in Preparing Honey. Right: Beehives Set near the Sagebrush, and the Especially Equipped Auto Truck in Which They are Moved About

Especially equipped Auto Truck

Eventually motors became commonplace on the farm. The automobile and truck changed the way people worked. One of the weak links in honey production was the need to uncap the honeycombs by hand. At the first beekeeping outfit I worked for in 1974, we ran two 45 frame extractors, pumped honey to the tank, etc. but we still used knives to cut the wax cappings away. These were electrically heated, a step above steam knives, but required skill and muscle to employ correctly. But in the 1800s, uncapping by knife was pretty much the only way to get the job done. In this vein, John H. Martin (aka, "The Rambler") gave a presentation to beekeepers at the California State Convention in the summer of 1900, titled "New Apiarian Inventions—Are They Needed?"

> We have examples in many portions of the country where the owners of many apiaries, and at least a thousand colonies of bees, are the ones that are deriving the greatest profit from them, and the profit is increased according as they adopt short cuts in the labor where hired help is dispensed with as far as possible. I will outline some work that is being done along this line with some degrees of success.

*When the exigencies of the time demand, we will have a machine
for uncapping the honey. I have gone so far with some experiments
in this line that I am quite sure that a machine can be constructed
that will uncap six or eight combs in just a few seconds; or, in other
words, you touch the button and the machine will do the rest.*

*I also certainly expect that the automobile will play an important part
in honey production. There is no bee-keeper who feels safe to drive
a span of horses near a bee-ranch, except in the night. (Martin 1900)*

Soon thereafter, patent applications for uncapping machines began to roll in. A.
C. Miller applied in 1901, and his device was issued a patent in 1902. It consisted
of two knives which would be moved back and forth with a foot operated treadle.
One of the problems devices of this type suffered from was the fact that many
beekeepers had frames of different width and the combs themselves varied in
thickness from one to two inches. By 1908, writers were still skeptical.

*Schemes for uncapping combs are still receiving more or less
attention. In the American Bee Journal for November is shown
a machine that works on a principle somewhat similar to some
described heretofore. But these machines have never gone beyond
the experimental stage. Possibly the application of steam heat to the
knives may go a long way toward solving the problem. (Anon 1908)*

Somewhat in the same vein, the writer declares:

*Strange as it may seem, a subscriber has a scheme for "milking" the
honey out of a hive without the removal of a comb or uncapping
the same; and, stranger still, he has actually sucked the honey out
of several supers by means of a strong suction-pipe. We are not at
liberty to give the details just now, but we expect to do so when the
inventor has perfected his invention. (Anon 1908)*

An article appeared in the 1909 *Gleanings in Bee Culture*, penned by R. F.
Holterman. He says in part:

*Various uncapping-machines have been suggested–two rollers,
for instance, with projecting pins to remove the cappings as they
revolve at high speed. I understand that a machine of this nature
has already been used, but it does not appear to have made much
headway. (Holterman 1909)*

Holterman goes on to praise the merits of the "Bayless Uncapping Machine." In the pictures it appears to be some sort of hand operated guillotine. The Editor comments extensively on the article and says:

> Whether Mr. Bayless is able to do more and better work with his machine than the average man can accomplish with the uncapping-knife, we can not say. For the present, at least, the public will doubtless be conservative, believing that, inasmuch as machine uncapping has proved a failure in the past, it will continue to do so. We shall see. (Holterman 1909)

I will end this section by referring to a 1924 manual published by A. I. Root Company titled "Extracting Honey." This little pamphlet shows a variety of large honey extractors, and discusses the various motor options, beginning with gasoline engines in 1907. Even in 1924, there is much discussion about the practicality of using electric power, due to the fact that electricity was not always available in rural locations. Further issues still remained as to the nature of the electric current available, whether low or high voltage, direct or alternating current. But most telling is that the only uncapping tool sold is the knife–steam heated, but still slicing combs by hand.

Comb Foundation

L. C. Dadant, writing in the 1924 *American Bee Journal* mused:

> There are a lot of things that have happened around our factory that are not recorded in our books, but are just as well remembered now as the day that the incidents happened. Probably my most vivid recollection of the early days of comb foundation was when, as a boy, I was loafing around the old log cabin which, at that time served as the foundation factory. In those days everything, of course, was done by hand, and the melting of beeswax was quite a task. (Dadant 1924)

The modern frame hive, the honey extractor, and comb foundation were three main inventions that were adopted in the second half of the 19th century. These items were mass produced by companies like Dadant & Sons and A. I. Root. Frames were made swiftly and accurately in factories but honey extractors were often constructed by beekeepers or local mechanics, being no more complicated than a bicycle. Comb foundation, on the other hand, was a bit tricky and while some attempted to make it with various contrivances that looked like waffle cookers, this product was best produced by specialists. By 1877, foundation mills were available but the making of beeswax sheets was a tedious process which involved dipping thin metal sheets in hot beeswax and peeling the wax away, to be fed in between engraved metal rollers.

Edward Beverly Weed, working with the A. I. Root Co, spent years on this problem and came up with a brilliant solution that quickly became the standard method of producing beeswax comb foundation. In the 1890s, Weed developed a rotating metal drum which had cold water circulating through it. This drum would slowly turn in a vat of hot melted beeswax which coated the outside of the drum. As it turned, the wax would be scraped off by a blade and forced through a slot, producing a continuous ribbon of wax of any width or thickness, depending on the size of the slot. These infinite ribbons would be taken up by a spool and cut when the rolled wax was about 18 inches in diameter. These would then be stacked to cool. Once cooled, the rolls would be mounted on a spool and fed through a pan of soapy water. Then the wax ribbon would be fed through the metal rollers which imprinted the pattern of the honeycomb into the wax. The wax could then be cut into uniform lengths.

Ferdinand Knorr
Making Candles

In 1975, I took a job at the Knorr Beeswax Factory in Del Mar, CA, where I learned how to run this system. On a good day, we would produce 1000 lbs. of comb foundation. The business had been started by Ferdinand Knorr, who fled Poland in 1904 and settled in Southern California. He kept bees but being a machinist, he saw an opportunity to make money selling comb foundation to the other beekeepers. By the time I worked there his son Henry, who was a top notch machinist himself, had taken over. He was equipped to refine one or two thousand pounds of crude beeswax daily, about half of which was made into candles. These were produced by a similar machine, which had the rotating drum that picked up hot wax, but then it was forced through dies and produced lengths of candle like a machine making huge macaroni. A skilled operator would cut these into six foot lengths. The cross section of the candle was like a wagon wheel, with spokes and a hole in the center into which the wicking would be pulled. Meanwhile in another part of the factory, wax rolls were being produced about 12 hours a day, as Henry worked sun-up to sundown.

As early as 1904, a letter to the magazine *Gleanings in Bee Culture* inquired about using sheet metal to reinforce the wax combs:

> *Can not very thin tin be used for foundation instead of wax, and a coating of wax be put on it? It would be much stronger, and not liable to break. I have to confess that I am not smart enough to extract the honey without breaking the comb. I used foundation the size of the frame, and wired it; but before I get up speed enough to throw the honey out, the comb is ruined by breaking.—A. H. Frank. Red House, N.Y. (Frank 1904)*

ONE HALF HORSE-POWER ELECTRIC MOTOR DRIVING FOUR-FRAME AUTOMATIC EXTRACTOR.

Electric Motor Driving Four-Frame Automatic Extractor

Shortly thereafter, a paragraph appeared in the *American Bee Journal* declaring:

> *Aluminum Honey-Comb—It is reported in Praktischer Wegweiser that combs are now made of aluminum, being no heavier than natural combs, combs which are promptly occupied by the bees for brood-rearing and storing. Some time ago metal combs were in use to a limited extent in this country, but we have heard nothing about them lately. The lightness of aluminum would seem to be a gain. (Anon 1911)*

For those who seek more information about the various wax comb replacements, I recommend "Substitutes for Beeswax in Comb and Comb Foundation" by T. S. K. Johansson & M. P. Johansson, in the journal *Bee World*.

Works Cited

Anon. (1911). "Aluminum Honey-Comb." *American Bee Journal*. Vol. 60, No. 3.

Dadant, L. C. (1924). "Recollections." *American Bee Journal*. Vol. 64, No. 12.

Frank, A. H. (1904). "Thin tin as a midrib for comb foundation." *Gleanings in Bee Culture*. Vol. 32, No. 19.

Hoffman, J. (1905). "Hoffman Frames." *Gleanings in Bee Culture*. Vol. 33, No. 11.

Holterman. R. F. (1909). "The Bayless Uncapping Machine." Gleanings in Bee Culture. Vol. 37, No. 6.

Johansson, T. S. K. & M. P. Johansson. "Substitutes for Beeswax in Comb and Comb Foundation." *Bee World*. Vol. 1, No. 4.

Martin, J. H. (1900). "New Apiarian Inventions—Are They Needed?" *American Bee Journal*. Vol. 40, No. 35.

ROLLING SHEETS TO OBTAIN CELL IMPRESSIONS.

Rolling sheets to obtain cell impressions

End note

Writing this book brought back lots of memories, especially of the stories Henry Knorr used to tell. He said when he was a boy, he hated bees because they used to extract the honey in a hot tent with a hand-cranker. This was placed right on the ground so in order to empty it, they dug a hole in the ground and ran the honey into a 5 gallon can. According to Henry they had no electricity or indoor plumbing: "It was like camping and I never cared for camping." Henry went through the whole transition, albeit in the late 1920s. He wanted to have nothing to do with bees, but when the opportunity came to take over the factory, that suited him fine. The shop is still there after 100 years. I would like to dedicate this book to the memory of Henry and Judy Knorr, their family and their many employees.